THE ROLE OF AGRICULTURE IN CLIMATE CHANGE MITIGATION

The Role of Agriculture in Climate Change Mitigation

Edited by:

Lucjan Pawłowski

Lublin University of Technology, Lublin, Poland

Zygmunt Litwińczuk

University of Life Sciences in Lublin, Poland

Guomo Zhou

Zhejiang A&F University, Hangzhou, China

CRC Press
Taylor & Francis Group
Boca Raton London New York

CRC Press is an imprint of the
Taylor & Francis Group, an **informa** business
A BALKEMA BOOK

CRC Press/Balkema is an imprint of the Taylor & Francis Group, an informa business

Typeset by Integra Software Services Pvt. Ltd., Pondicherry, India

Library of Congress Cataloging-in-Publication Data

Applied for

Published by: CRC Press/Balkema
Schipholweg 107C, 2316XC Leiden, The Netherlands
e-mail: jose.vanderveer@taylorandfrancis.com
www.crcpress.com – www.taylorandfrancis.com

ISBN: 978-0-367-43372-7 (Hbk)
ISBN: 978-1-003-00273-4 (eBook)
DOI: 10.1201/9781003002734
DOI: https://doi.org/10.1201/9781003002734

The Role of Agriculture in Climate Change Mitigation – Pawłowski, Litwińczuk & Zhou (eds)
© 2020 Taylor & Francis Group, London, ISBN 978-0-367-43372-7

Table of contents

 ISBN 978-0-367-43372-7

Preface

According to the IPCC reports, one of the greatest threats to the Earth ecosystems is the climate change caused by the anthropogenic emissions of greenhouse gases, mostly carbon dioxide. The greenhouse effect itself is, to a certain level, a beneficial phenomenon, because if the heat radiated from the surface of the Earth were not retained by the greenhouse gases and water vapour, the mean temperature on the surface of the Earth would be approximately -14°C. The problem is the excessive emissions of greenhouse gases from anthropogenic sources, mainly from the combustion of fossil fuels, cement production and land-use change which lead to excessive temperature rise.

According to the authors of the IPCC reports, the consequences of further temperature rise will be catastrophic for the Earth's ecosystems. However, a less catastrophic opinion should be noted, for instance, R. Lindzen – an outstanding American climatologist – does not negate the existence of greenhouse effect; however, he expects that the magnitude of climate change will be far milder (Lindzen 1990, Lindzen 2010).

An excessively one-sided approach to mitigating the CO_2 emission from anthropogenic sources, focusing on reducing fossil fuel combustion, may slow down the economic development of many countries. Generation of energy from renewable sources sometimes leads to the creation of socio-economic problems (Pawłowski et al. 2017, Bucher 2016, Czyżewski et al. 2018, Krajewski 2019, Olejnik et al. 2017). The negative examples include the production of biodiesel fuel from the coconut palm oil grown in Indonesia on the land acquired by burning off tropical forests. Promotion of biofuels was based on a simplified analysis and the assumption that the amount of CO_2 produced during the biofuels combustion is equal to the amount absorbed from the atmosphere in photosynthesis (Pawłowski et al. 2017).

Although this statement is true, it does not account for the additional energy use connected with the cultivation, harvest, and processing the plants into biofuel. Moreover, such assumption omits the fact that in order to create a plantation, another ecosystem was destroyed – such as tropical forest or peatland – which would absorb greater amounts of CO_2 from the atmosphere. Therefore, introducing renewable energy sources requires an in-depth analysis of both socio-economic and environmental effects (Olejnik et al. 2017, Kumar et al. 2019, Ciais et al. 2013, Bułkowska et al. 2016, Pawłowska et al. 2017, Montusiewicz et al. 2008, Adhikari et al. 2015, Richardson et al. 2015, Molino et al. 2018, Le Quere et al. 2018,Toochi 2018).

Brazil is a country which successfully introduced ethanol for fuelling cars on a large scale. A comprehensive programme of utilizing sugar cane for ethanol production was developed, which also took into consideration the socio-economic and environmental conditions (Bucher 2016).

While the use of plants as a source of energy is highly recommended for mitigation of CO_2 emissions, the role of terrestrial ecosystems, including agriculture (Ciais et al. 2013) is underestimated.

In the Earth ecosystems, there are four natural CO_2 fluxes (Le Quere 2018):

- absorption of CO_2 by plan in the amount of 451 $GtCO_2$/yr
- emissions (respiration) of CO_2 in the amount of 435 Gt CO_2/yr
- absorption of CO_2 in oceans in the amount of 293 Gt CO_2/yr
- desorption of CO_2 from degassing of oceans in the amount of 287 Gt CO_2/yr)

It means that all Earth ecosystems absorb 744 $GtCO_2$/yr and simultaneously emit 722 Gt CO_2/yr.

Additionally to these natural CO_2 fluxes, there are two anthropogenic CO_2 fluxes:

- emissions from industrial activity (E_{ff}) i.e. from the combustion of fossil fuel and cement production, and
- emissions from land-use change (E_{LUC})

As it is shown in Table 1, global emissions were growing from 16.9 Gt CO_2/yr during the period of 1960-1969 to 42.2 Gt CO_2/yr in 2018 (Le Quere 2018). In case of Poland total CO_2 emission grows from 299.4 mln ton CO_2/yr to 322.5 mln ton CO_2/yr in 2018 (Table 2) One possibility of neutralizing these emissions of CO_2 is increase in CO_2 absorption by land ecosystem by enhancement of biomass grow through fertilization and better utilization of land i.e. an increase intercrop cultivation which leads to a significant increase in the CO_2 absorption by the photosynthesis process (Ciais 2013). An additional biomass produced can be used as a green fertilizer and raw material for energy generation. Using the intercrop biomass as green fertilizer will decrease the use of fossil fuels for the production of synthetic fertilizer, and thus decrease the CO_2 emissions from these sources.

Table 1. Characteristic of global CO_2 emissions ($GtCO_2$/yr) from fossil fuel combustion and cement production (E_{ff}) and land use from change (E_{LUC}).

years emissions	1960-1969	1970-1979	1980-1989	1990-1999	2000-2009	2008-2017	2017	2018
E_{ff}	11.4	17.2	20.2	23.5	28.6	34.5	36.3	37.1
E_{LUC}	5.5	4.8	5.1	5.9	3.7	5.5	5.1	5.1
Total	16.9	22.0	25.3	29.4	32.3	40.0	41.4	42.2

Table 2. Emissions of CO_2 in Poland 2000-2018.

2000	2001	2002	2003	2004	2005
299.4	297.7	294.2	303.4	301.4	307
2006	2007	2008	2009	2010	2011
320.2	317	319.2	305	322.8	322.6
2012	2013	2014	2015	2016	2017
307.2	304.8	292.9	292.9	305.6	315.4
2018					
322.5					

The intercrop biomass can be used directly as a fuel or after conversion to biogas or ethanol in the processes of fermentation or conversion to syngas and biochar in the pyrolysis process (Molino 2018).

Taking into account large natural CO_2 fluxes in the process of photosynthesis (451 $GtCO_2$/yr) it is easy to find that the increase of photosynthesis process by 9,7% would naturalize all anthropogenic CO_2 emissions (42.2 $GtCO_2$/yr). There is a widespread belief that forest as CO_2 sink could significantly contribute to the mitigation of climate change (Toochi 2018).

However, it is less known greenhouse gases emissions by forest. First of all, forest emits carbon dioxide during respiration and decomposition of litters. Less known is the emission of methane, nitrous oxide and terpenes which are strong greenhouse gases. Additionally, forest due to dark colour absorbs more sunlight and thus heats the earth's surface more.

To evaluate the impact of agriculture on climate change, one should take into account ruminant farming. These animals emit considerable amounts of methane which has strong greenhouse effects. This methane emissions may be reduced by using appropriate feed for ruminants. Some researchers also postulate to reduce meat consumption from these animals.

Generally speaking, agriculture and forestry are responsible for quiet big emissions of greenhouse gases and as well as a great potential for their reduction.

REFERENCES

Adhikari, U., Eikeland, S., Halvorsen, B. 2015. Gasification of biomass for production of syngas for biofuel, Proceeding of the 56th SIMS, October 07-09, Linköping, Sweden.

Bucher, S. 2016. Sustainable Development in the World from Health and Human Development Index: Regional Variations and Patterns, *Problemy Ekorozwoju/Problems Sustainable Development* 12(1):117–124.

Bułkowska, K., Gusiatin Z.M, Klimiuk E., Pawłowski A., Pokój T., *Biomass for Biofuels*, 2016, CRC Press.

Ciais, P., et al. 2013. 5th IPCC Report 2013, Carbon and other Biogeochemical Cycles.

Czyżewski, B., Guth, M., Matuszczak, A. 2018. The impact of the CAP Green Programmes on farm productivity and its social contribution, *Problemy Ekorozwoju/ Problems Sustainable Development* 13(1):173–183.

Krajewski, P. 2019. Bioeconomy – opportunities and dilemmas in the context of human rights protection and environmental resource management, *Problemy Ekorozwoju/ Problems Sustainable Development* 14(2):71–79.

Kumar, P., Sharma, H., Pal, N., Sadhu, K. 2019. Comparative assessment and obstacles in the advancement of renewable energy in India and China, *Problemy Ekorozwoju/ Problems Sustainable Development* 14(1):173–183.

Le Quere et al. 2018. Global Carbon Budgest 2018, *Earth Syst. Sci. Data*, 10, 2141–2194.

Lindzen, R. 1990. Dynamic in Atmospheric Physics. Cambridge University Press, DOI: https://doi.org/10.1017/CBO9780511608285

Lindzen, R. 2010. Global Warming: The Origin and Nature of the Alleged Scientific consensus. *Problemy Ekorozwoju/Problems Sustainable Development* 5(2):13–28.

Molino A., Larocca V., Chianese S., Musmarra D. 2018. Biofuels production by biomass gasification: A review, *Energies*, 11:811, Doi: 10.3390/en11040811

Montusiewicz, A., Lebiocka, M., Pawłowska, M. 2008. Characterization of the biomethanization processes in selected waste mixtures. *Archives Environmentl Protection* 34(3):49–61.

Olejnik, T., Sobiecka, E. 2017. Utilitarian technological solutions to reduce CO_2 emission in the aspect of sustainable development, *Problemy Ekorozwoju/Problems Sustainable Development* 12(2):191–200.

Pawłowska, M., Pawłowski, A. 2017. Advances in Renewable Energy Research., Taylor and Francis Group, London, UK 2017, ISBN: 978-1-138-55367-5 (Hbk), ISBN: 978-1-315-14867-0 (eBook).

Pawłowski, A. & Pawłowska, M. 2017. Mitigation of greenhouse gases emissions by management of land ecosystems. *Ecological Chemistry and Engineerin* 24(2):213–221.

Pawłowski L. 2012. Do the liberal capitalism and globalization enable the implementation of sustainable development strategy?, *Problemy Ekorozwoju/Problems Sustainable Development* 7(2): 7–13.

Richardson, Y., Drobek, M., Julbe, A., Blin J., Pinta, F. 2015. Chapter 8 – Biomass gasification to produce syngas, Recent Advances in Thermo-Chemical Conversion of Biomass, 213–250.

Toochi, E.C. 2018. Carbon sequestration; how much can forest sequestrate CO_2? *Forest Res. Eng. Int. J.* 2(3):148–150.

 ISBN 978-0-367-43372-7

About the editors

Lucjan Pawłowski

Lucjan Pawłowski, born in Poland in 1946, is the Member of the Polish Academy of Science, Member of the European Academy of Science and Arts, honorary professor of China Academy of Science, President of Environmental Engineering Committee of the Polish Academy of Science. Director of the Institute of Environmental Protection Engineering of the Lublin University of Technology. He got his Ph.D. in 1976, and D.Sc. (habilitation) in 1980 both at the Wrocław University of Technology). He started research on the application of ion exchange for water and wastewater treatment. As a result he together with B. Bolto from CSIRO Australia, has published a book "Wastewater Treatment by Ion Exchange" in which they summarized their own results and experience of the ion exchange area. He has published 22 books, over 168 papers, and authored 108 patents, and is a member of the editorial board of numerous international and national scientific and technical journals.

Zygmunt Litwińczuk

Zygmunt Litwińczuk, born in Poland, is the Rector of Lublin University of Life Sciences and the head of the Department of Cattle Breeding and Protection of Genetic Resources.
His scientific interests are concentrated mainly on the protection of genetic resources of indigenous cattle breeds (Polish Red, Whiteback) and controlling of emission from ruminants. Professor Litwińczuk published over 700 publications, including over 300 original works and more than 20 monographs and handbooks.

Guomo Zhou

Guomo Zhou, Ph.D., was born in China, 1961. He received his Ph.D. from Zhejiang University in 2006 and served as the president of Zhejiang A&F University from 2008 to 2017. Currently he is the chancellor of the Zhejiang A&F University, director of the State Key Laboratory of Subtropical Silviculture, and honorary professor of Lublin University of Technology of Poland. He systematically clarified the scientific issue of whether bamboo forest is a carbon source or a carbon sink, which resolved the international dispute on the source and sink, and allowed the bamboo forest to be included in the international forest emission reduction category. He has developed an ecosystem management technology for increasing bamboo forest carbon stocks, reducing carbon emission and stabilizing soil carbon stocks. The technology was listed as "the national key promotion of low carbon technology". He has developed two methodologies for the bamboo forest carbon sequestration project which solved the technical bottleneck of bamboo forest carbon sequestration project when entering the carbon emission reduction market. He has published 103 papers in journals indexed with SCI, making him the worldwide highest productive and most influenced author in the area of bamboo carbon sequestration. As an expert on forest carbon sequestration, he has given keynotes at the side event of the UNFCCC/COP for six times and submitted two technical reports since 2009 in Copenhagen and served as an evaluator for the IPCC Assessment Report for Climate Change.

The Role of Agriculture in Climate Change Mitigation – Pawłowski, Litwińczuk & Zhou (eds)
© 2020 Taylor & Francis Group, London, ISBN 978-0-367-43372-7

Sustainable energy: Technology, industry, transport and agriculture

A. Pawłowski
Faculty of Environmental Engineering, Lublin University of Technology, Lublin, Poland

1 INTRODUCTION

According to the most widely used definition sustainable development it is development that meets the needs of the present without compromising the ability of future generations to meet their own needs (WCED 1987). It includes basic human needs, like access to water, food and shelter, but also to electricity.

Sustainable development is discussed at three levels, i.e. ecological, social, and economic (Borys 2005, Harris et. al. 2001, Singh 2019, Sztumski 2019). In my works I proposed that the discussion should be expanded to include other dimensions. When addressing the multi-dimensional nature of sustainable development, I also pointed to (Pawłowski 2011):

- ✓ the ethical dimension (the question of human responsibility for nature);
- ✓ the ecological dimension (nature conservation, protection of man-created environment, spatial planning);
- ✓ the social dimension (not only the natural environment but also the social environment may undergo degradation);
- ✓ the economic dimension (taxes, grants and other economic instruments);
- ✓ the technical and technological dimension (new technologies, economical use of raw materials);
- ✓ the legal dimension (environmental law);
- ✓ the political dimension (formulation, implementation and enforcement of sustainable development strategies).

Above dimensions may be presented in a hierarchical order (Table 1).

On the first level I place ethical reflection, which provides a basis for the other levels. It is evident, that when a person takes a decision in line with his/her own convictions and system of values, the situation is quite different from the one when a decision is the result of obligations and prohibitions of the law currently in force.

What provides the foundation for the whole debate is the ethical justification for valid questions about the values to be adopted and the reasons to proceed in one way and not another.

The second level comprises matters ecological, social and economic, which are accorded the same weight.

The third level contains an analysis of detailed technical, legal and political issues.

The traditional discussion on sustainable development has centred around the second level. Yet unless it is rooted in ethics (the first level), it will remain incomplete. Without the third level, in turn, precise practical solutions will prove elusive.

The proposed hierarchy offers a new perspective on sustainable development. The wide scope of issues that is proposed – together with the equally wide-ranging changes proposed for the different planes and the strategies already adopted – allows us to formulate the following claim: should it actually be implemented, sustainable development will introduce a new order as revolutionary as the breakthroughs in human history that are conventionally dubbed Revolutions (Pawłowski 2011).

Table 1. The hierarchy of levels of sustainable development (author's own work).

Level I		Ethical dimension	
Level II	Ecological dimension	Social dimension	Economic dimension
Level III	Technical Dimension	Legal dimension	Political dimension

Table 2. Top 10 of total energy consumption by countries in 2018 (Global Energy Statistical Yearbook, 2018).

No	Country	Energy consumption [Mtoe]
1	China	3105
2	United States of America	2201
3	India	934
4	Russia	744
5	Japan	429
6	Germany	314
7	South Korea	296
8	Brazil	291
9	Canada	287
10	Iran	253

2 THE NEED FOR ENERGY

One of the most important challenges of the Sustainable Development Revolution is connected with energy (Żelazna & Gołębiowska 2015, Kumar et al. 2019, Johansson & Pirouzfar 2019). According to UNDP, between 2000 and 2016 the number of people with electricity increased from 78 to 87 %. Also the demand for energy is growing almost everywhere, especially in Asia. In 2040 it will double comparing to the year 2010. Our civilization cannot sustain itself without electricity. The biggest consumers of energy are presented in Table 2.

It is worth to mention the example of famous Itaipu hydropower station (Wikipedia, 2019). It was built in the years 1975-1984 and produces 95 TWh yearly. It covers 95% of electricity demand in Paraguay and 20% in Brazil. On January 21st 2001 as much as 13 of the 18 turbines stopped, because of the broken high voltage line. Metropolis like São Paulo (populated by 12 million people) and Rio de Janeiro (populated by 6,5 million people) for few days had to live without electricity. Similar disaster happened on November 11th 2009. In both cases it was a total blackout and total chaos. So we need electricity, and the problem is not only how to produce the amount of energy that we need, but also in which way to produce it, since there are different ways, with different environmental consequences (Gore 2017).

In general we may say that the electricity may be produced from:

- fossil fuels,
- nuclear power,
- renewable sources of energy.

The situation is changing, as it is presented on Figure 1 and 2.

First picture is presenting the situation from the year 2005. As we can see, 68.5% of energy was produced from fossil fuels, however the second place was already for one of the renewables – hydropower, 19.7%. Than nuclear power (9,65%) and renewables other than hydropower (only 2.15%).

Second figure is from the year 2015. It looks similar, however there are some important changes. First place is still for fossil fuels, but the percentage is lower (64.97%) and second place is again for hydropower (20.02%). But there is a shift between next two places. Renewables other than hydropower are gaining power (9.65%) and last place is for nuclear power (6.73%).

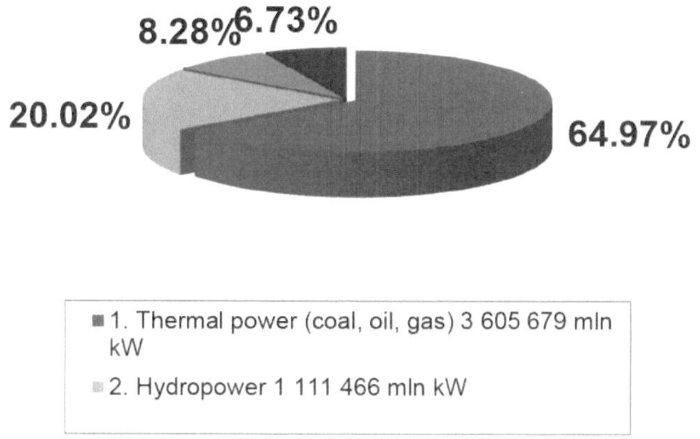

Figure 1. The world's perspective: installed electricity capacity by type (International Energy Annual, Washington D.C. 2005).

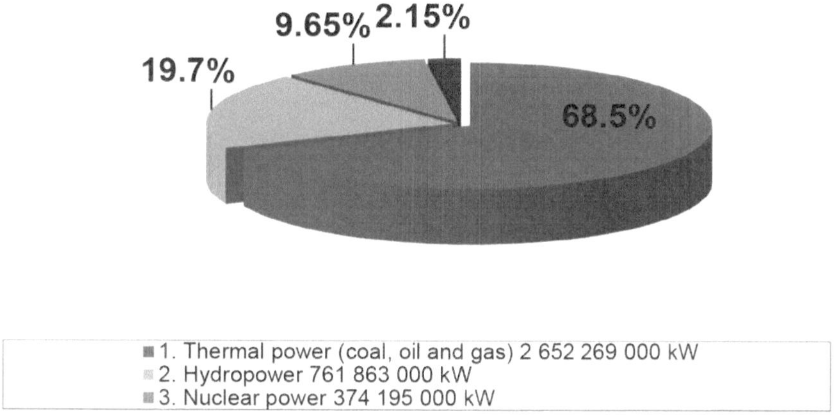

Figure 2. The world's perspective: installed electricity capacity by type (International Energy Annual, Washington D.C., 2015).

There at least two reasons for this.

In case of nuclear power the accident in Fukushima in 2011 showed, that always may happen something unexpected.

In case of renewables there are regions, where such sources of energy are getting strong support, like in the European Union.

Let's look at the problems connected with energy from the perspective of sustainable development. The term sustainable energy has recently gained great popularity (E&C Consultants 2019). It is one which is able to meet the growing demand of today's people without compromising the demand of the people that would require it in future (CEF 2019).

This energy:

- Supposed to have minimal impact in the natural environment.
- Can be naturally replenished.
- Is inexhaustible, so there is no depletion of resource over time.
- Its efficiency should improve over time (CEF 2019, United Nations Global Compact 2012).

In many scientific papers sustainable energy means using of renewable sources of energy. However, it is not always so, since renewables may be also dangerous to the environment, and coal is not always so bad.

3 THE FUTURE OF COAL

Fossil fuels are perceived as unsustainable source of energy, mainly due to large emissions of carbon dioxide, the most important gas contributing to the greenhouse effect. Energy sector is responsible for around 60% of the total anthropogenic carbon dioxide emissions (Pawłowski 2011).

The greenhouse effect is connected with the concentration of the greenhouse gases (GHG) in the atmosphere. Short-wave solar radiation, that heats up the Earth, penetrates through GHGs. The heat so generated (long-wave radiation) will be to a large extent reflected by the GHG layer back towards the Earth's surface (this phenomenon is known as back-radiation). The mechanism is the same as in a traditional greenhouse – where glass panes have the same effect as greenhouse gases.

Greenhouse gases include (EPA 2019):

- carbon dioxide (50% share in the greenhouse effect),
- methane (18%),
- freons (14%),
- ozone (12%),
- nitrogen oxides (6%).

We should add to this water vapour – which is also a GHG and its content in the atmosphere increases along with the increase in global temperature (O'Neill 1997).

50% of the greenhouse effect is attributed to carbon dioxide; therefore, it would be worthwhile to present countries that are responsible for the highest largest CO_2 emissions (Drastichová 2017).

In general, global CO_2 emissions are growing and in 2017 reached 37,077.404 Mt, of which most was generated by the China, USA and the EU. Unfortunately, in case of the most of 10 countries with the highest emission of carbon dioxide the emission is still growing (Table 2).

Excessive emissions of GHGs from anthropogenic sources may melt the polar caps, cause flooding of the vast coastal zones of seas and oceans, or increase the area affected by climate anomalies. Other problems include a decline in agricultural production (due to the increased population of the wintering pests) as well as certain diseases (fungal, bacterial, and viral), and seed germination problems caused by the rising temperature.

The share of anthropogenic gases in the total GHG pool, as well as the impact of other factors on climate changes, are still a matter of discussion (Lindzen 2010). These other factors include e.g. deforestation, and in particular the destruction of rainforests. It is common

Table 2. Top 10 countries with the highest emission of carbon dioxide in 2017 (Wikipedia).

No	Country	Emission of CO_2 in Mt per year	Trend in emission level
1	China	10,877.218	▲
2	USA	5,107.393	▼
3	European Union	3,548.345	▼
4	India	2,454.774	▲
5	Russia	1,764.866	▼
6	Japan	1,320.776	▲
7	Germany	796.529	▼
8	South Korea	673.324	▲
9	Iran	671.450	▲
10	Saudi Arabia	638.762	▲

knowledge that during the 20th century, as much as 50% of the rainforests were destroyed due to deforestation, still about 12 million hectares are removed every single year, and about 50 hectares are destroyed every minute (Kalinowska 1992). As forests are known for their climate stabilizing function, removal of such a vast portion of their global resources must have a crucial destabilizing impact on the Earth's climate!

So what about coal?

First of all there are new technologies, which are connected with coal, and much better to the environment than traditional solutions. They are called Clean Coal Technologies (CCT). Among them let me discuss two such technologies: Carbon Capture and Storage (CCS) and Blue Coal.

Carbon Capture and Storage (CSS) is an American technology. In this case there is almost no emission of carbon dioxide to the atmosphere, the reduction reaches the level of 80-90% compared to traditional coal burning plant. Instead of the emission to the atmosphere this gas is stored under high pressure below the Earth's surface in geological formations, for example in depleted gas reservoirs, or in the form of mineral carbonates. In 2019 there are 17 such systems introduced in the world, being able to capture 31,5 Mt of carbon dioxide every year. In case of United States, according to the National Energy Technology Laboratory, it is believed that possible storage capacity is enough for the next 900 years, taking into account the present level of electricity production from coal (Phelb et al. 2015).

Blue Coal is a Polish technology (Piłat 2015). It is a specially prepared coal that has been partially degassed at 45000C. It contains significantly less impurities, burns smokeless and still has a high calorific value!

Secondly there is a known phenomenon called sequestration of carbon dioxide by terrestrial ecosystems. It will be discussed in a section connected with agriculture.

If we add to this known fact, that proven reserves of coal will last at least for the next 120 years (Pawłowski 2011), the total resignation from the use of coal seems to be irrational. However for sure we must work on other sources of energy, to get more balanced mix of energy.

4 THE FUTURE OF NUCLEAR POWER

Nuclear power is an alternative. It is not renewable source of energy, however the reserves of uranium are very large, they will last for at least 130 years, however with new technologies, like nuclear fusion reactors, this time may be expanded to even 1000 years (Pawłowski 2011).

In 2017 the countries with the biggest production of energy from nuclear power were USA (805 647 GWh) and France (381 846 GWh). In Europe, nuclear power plants are located also in Finland, Germany, Russia and Ukraine (Salay 1997, Pawłowski 2011).

There are two types of reactors are mainly used in the world today (Pawłowski 2011):

- PWR – Pressurized Water Reactors. They generate heat carried by water to a steam generator. Water is transported under high pressure to prevent water boiling in the reactor's cooling system. This technology accounts for 60% of the market.
- BWR – Boiling Water Reactors. Water plays a key role also in this type of reactors. Steam is generated in the reactor and used to drive a turbine. This technology accounts for 24% of the market.

The main problems in this area are as follows (Pawłowski 2011):

- Technical safety of the power plant and the undefined costs of closing of the nuclear power units after the power plant is de-commissioned.
- Management of the radioactive waste – the proposed waste management solutions (storing waste in closed underground mines, drilling of underground tunnels, or even ocean floor disposal) seem to be insufficiently safe.
- Risks related to nuclear tests and the potential use of nuclear weapons during armed conflicts.
- Development of nuclear technologies in the Third World countries.

The first controlled nuclear chain reaction took place in the world's first nuclear reactor in Chicago, the USA, in 1942 (Universal Encyclopedia 2000).

Three years later, atomic bombs were dropped on Hiroshima and Nagasaki (6 August and 9 August 1945), marking a turning point in World War II, and at the same time causing an unprecedented environmental and social disaster.

Soon, the arms race between the West and the East began. As a result, nuclear power plants were built not only to generate electricity, but also to produce materials to be used in atomic bomb manufacture.

Technological safety proved to be the key element.

The explosion in unit 4 in the power plant in Chernobyl on 26 April 1986 (Jackson 1996) and melting of the nuclear core was to a large extent related to the type of reactors used (RBMK or channel-type graphite-moderated reactors).

The same problem (core melt-down) was recorded on 29 March 1979 in Pennsylvania in the Three Mile Island nuclear power plant, but the installed safety systems prevented an explosion in this case (The Need for... 1979).

The main problem in RBMK reactors is the use of graphite as the neutron moderator. In other types of reactors, water is used as a moderator, and in the event of a water supply system breakdown, the chain reaction is stopped. In RBMKs the chain reaction is accelerated after the water supply is closed, leading to a disaster. As a result, this type of reactor was banned in many countries, including the United Kingdom (after a dangerous graphite stack fire in an RBMK reactor in an experimental stage).

In the Soviet Union, RBMKs were still built. The reason was twofold: both economic and military.

First of all, RBMKs can use low-enriched uranium, which is much cheaper. And secondly, RBMKs generate large amounts of plutonium that can be used to produce nuclear warheads.

The exploding reactor in Chernobyl emitted large amounts of radioactive materials to the air – 50,000,000 Ci according to official sources (equivalent of about 500 bombs dropped on Hiroshima), and 300,000,000 Ci according to unofficial sources (equivalent to 3000 bombs). In addition, the area within the distance of 30 km around the reactor was closed, and 30,000 people were evacuated and resettled (Jaśkowski 1988).

However, it was not the only such serious nuclear disaster in history. In 1957, a tank holding 160 tonnes of highly radioactive waste from nuclear warheads exploded in the Kyshtym plant in Chelyabinsk in the Ural Mountains. The plant was located away from the European part of the Soviet Union, which made it possible to keep the disaster secret.

It was officially revealed 40 years later. The area of 15,000 square kilometers was affected, of which at least 20% is still classified as highly radioactive (Borkowski 1991).

The use of nuclear power in submarines is another problem.

Several submarines are known to have sunk as a result of a nuclear breakdown.

The most famous being the Russian submarine Kursk that sank on 12 August 2000. The wreck was cut into pieces and recovered during two operations in 2001 and 2002.

Another example is the case of the Komsomolets submarine that sank on 7 April 1989 in the North Sea while carrying nuclear missiles. In 1994, leaks of plutonium had been reported. Work to improve the tightness of its hull were undertaken in 1995-1996. The question remains: is it possible that radioactive substances will be ultimately released to the environment, causing nuclear pollution of the sea?

Even the most state-of-the-art and peaceful technologies still carry certain risks.

In September 1999, a serious near-miss situation occurred in the Japanese power plant of Takaimura due to the non-compliance with safety procedures.

So, no technology – but people were the reason. The quantity of enriched uranium added to a container exceeded the technical standards – by as much as seven times. A chain reaction was started. Luckily, it was stopped just in time. However, the increase in radiation levels was high enough to cause two deaths, and several hundred other staff members were exposed to excessive radiation (New Details on.., 1999). This accident resulted in a number of critical publications in the press, discussing the safety of cutting-edge technologies used in Japanese nuclear power plants (Edwards, 2002).

Then the accident in Fukushima in 2011. They had many security installations, but did not predict that water pumping pumps may be flooded, which meant no cooling for the reactor and disaster for the power plant.

In case of nuclear power the accident in Fukushima in 2011 showed, that always may happen something unexpected. They had many security installations, but did not predict that water pumping pumps may be flooded, which meant no cooling for the reactor which means disaster for the power plant.

The safety of European reactors is also questioned. In 2002, a reactor in Denmark was closed due to safety concerns (Vince 2002).

These aspects should be considered when analyzing the perspectives for the future development of nuclear power (Moberg 1986). As for now, the nuclear energy market is slowly shrinking, however in some areas this source will be important for many years to come (Table 3).

Among countries with the highest production of electricity from nuclear power we must mention Japan, South Korea, China, Canada, Ukraine, France, USA, Russia, and Germany. I think that there are some more perspectives in Europe. There is strong pressure to stop burning coal in the EU and using more renewables. However, mainly due to climate conditions, renewables will never get the 100% share in the market in this region, so if not coal, nuclear power is the alternative.

5 THE PERSPECTIVES OF USING RENEWABLE SOURCES OF ENERGY

It is worth to mention the legislation of the European Union, as the most ambitious project from the world's perspective.

First program was called 3 x 20 to the year 2020, and its basic principles were as follows (EC 2009):

- 20 % better energy efficiency,
- 20 % less of emission of the greenhouse gases,
- 20 % of energy of the member states from renewables,

Next new horizon was established, covering the time to the year 2030, with new goals (EC, 2018):

- further improvement in energy efficiency, at least to 32,5%
- further reduction of the greenhouse gases emissions (40% to the year 2030, in relation to the level of emissions from 1990),
- 27% of energy of the member states from renewables.

Table 3. Top 10 of nuclear power plants with the highest energy production (Wikipedia).

No	Name of the plant	Country	Capacity MW
1	Kashiwazaki-Kariwa	Japan	7965*
2	Kori	South Korea	7411
3	Yangijang	China	6480
4	Bruce	Canada	6384
5	Hanul	South Korea	5928
6	Hanbit	South Korea	5875
7	Zaporizhia	Ukraine	5700
8	Gravelines	France	5460
9	Paluel	France	5320
10	Cattenom	France	5200

* temporary out of service

Also United Nations legislation is supporting renewables. Among 17 Sustainable Development Goals, introduced by the UN in 2015, one is especially important: Goal no 7: Affordable and Clean Energy, which means not only better energy efficiency but also using as much as possible of renewables. Few other goals are interconnected with this, like Goal no 9: Industry, Innovation and Infrastructure, as well as Goal no 8: Decent work and Economic Growth. Clean Renewable Energy is a huge part of the market, which means it is giving a lot of work for millions of people. According to the UNDP, the renewable energy sector in 2017 gave employment to as much as 10.3 million people. Renewables need also new infrastructure, which is in compliance with goal no 9.

Renewable energy sources include (Pawłowski, 2011):

- biomass,
- hydropower,
- solar energy,
- wind power,
- geothermal energy,
- biogas.

Table 4 presents share of different renewables in the global market.

In the case of biomass and biogas burning, as well as geothermal energy sources, emissions to the environment are generated (mainly NO_2 and SO_2 for biomass and biogas, and CO_2 and hydrogen sulphide for geothermal power), but their level is much lower than in the case of fossil fuels (Boyle 1996, Cao & Pawłowski 2013, Bułkowska et al. 2016). For other renewable energy sources (RES), emissions are close to zero, although the equipment used to generate RES power is manufactured using metals and technologies that are not environmentally neutral (Velkin & Shcheklein 2017).

Let us analyze the general conditions for the use of the environmentally friendly renewable energy sources.

5.1 *Biomass burning*

Biomass materials include straw, peat, wood waste (from agriculture, forestry, or paper industry), as well as timber from special plantations (Pawłowski 2011). In Europe, Osier Willow (salix viminalis) is a popular source of biomass. Other sources include poplar (populus), Black Locust (robinia pseudacacia) and rose (rosa multiflora).

The characteristic features of these species include (Mackow et. al. 1993):

- fast growth, even 10 times faster than the natural growth of biomass in an ordinary forest,
- high calorific value: two tonnes of dry timber equal one tonne of coal,
- no special requirements as regards soil and climate conditions – fallow land, land set aside, or even polluted or flooded land can be used,
- no special requirements as to temperature and climate – resistant to low temperatures,
- high resistance to pests and diseases,

Table 4. Renewable electricity generation by energy use (IRENA 2018).

No	Source of energy	Electricity generated [TWh]	Share in the market [%]
1	Hydropower	4 049	69%
2	Wind energy	958	16%
3	Bioenergy*	467	8%
4	Solar energy	83	6%
5	Geothermal energy	1	1%

* Bioenergy includes solid biofuels, biogas, renewable municipal waste, liquid biofuels

- extended use of plantation (in the case of willows, crops are harvested every 2-3 years and the plantation may be used for up to 25 years).

However, there are some problems:

- The dynamic development of agricultural crops for biofuels in the European Union countries caused in the years 2007-2010 a dramatic 2.5-fold increase in the food prices index (according to the United Nations Food and Agriculture Organization).
- In addition, the use of biofuels for transport caused, that especially in developing countries, the tropical forests are vanishing more rapidly than before and replaced with biofuel plants. It has negative effect on the climate, since forests play important role in stabilization of the climate. What's more, converting natural ecosystems to produce energy plants in itself is responsible for huge emissions of CO_2, even 400 times more than using such source of energy will save a year!
- The problem is also with transporting biomass over longer distances, even between the continents. In case of such practices energy efficiency is dramatically falling!

So what about the future of biomass? The problems do exist, but we are aware of them and all of them are possible to overcome. The most important issue is that biomass is everywhere. And if we compare the available potential with the present use, we will find out that only 2/5 of biomass resources are in use (Vlosky & Smithhart, 2011), so this source of energy has the future.

5.2 *Hydropower plants*

The main task of hydropower plants is to produce electricity. There is no need to produce steam as in coal-fired and nuclear power plants – hydropower itself drives the turbine that supplies energy to the power generator (Pawłowski 2011).

From the technical point of view, there are four basic types of hydropower plants (Kucowski, Laudyn & Przekwas 1993):

- run-of-the-river plants (without a dam),
- conventional plants (with a dam and reservoir),
- pumped storage plants (reservoir is located above the plant; when the demand for energy is lower, for instance at night, the plant will pump water back to the reservoir to replenish the storage of water),
- tidal power plants: using the power of tides to drive the turbine in both directions.

Also the OTEC plants (ocean thermal energy conversion plants) are an interesting solution. They are located in oceans near the equator, in the tropics, in Hawaii, Japan, or Indonesia. In this zone, the temperature of water is about 300C at the surface, but only 70C at the depth of 300-500 m. This difference is used as a source of heat converted to electricity.

The largest hydropower plant in the world is in China – the Three Gorges Dam on the Yangtze River. The plant was opened on 24 June 2003, but the full capacity of its storage reservoir was achieved in October 2010 (39.3 billion cubic meters, the reservoir stretches 700 km along the Mudong River). In 2008 it generated 80,8 TWh, and together with Gezhouba Dam the complex generated 97,9 TWh (Wikipedia 2018).

A second plant of this type is the Itaipu hydropower plant on the Parana river on the Brazil – Paraguay border, which in the same year generated 94,68 TWh (Wikipedia, 2018).

In many other countries there are no such powerful rivers, but even small hydropower plants (below 10 or 5 MW) offer a number of benefits such as the following:

- electricity is produced without causing any pollution,
- protection by dams against flooding is provided,
- it is stable source of energy in comparison with some other renewable sources of energy, like wind or sun.

There are also a number of environmental impacts that give rise to serious reservations:

- change of the local environmental conditions (especially hydrological), the landscape, or even the climate, which is an obstacle in the case of projects located near natural areas, like national parks, protected by law,
- obstacles to the natural migrations of fish,
- necessary relocation of inhabitants of human settlements that will be covered with water (in the case of the Three Gorges Dam on the Yangtze River, as many as 1.4 million people had to be relocated),
- problems keeping the optimum purity of water in storage reservoirs.

Therefore, the proper location of the dam is extremely important. Not everything may be however predicted. Droughts are due to climate changes more and more often which can cause especially small rivers to dry out. This may stop the development of small hydropower plants.

5.3 *Solar energy*

Solar energy is necessary for life on Earth. Living organisms are able to use only about 0.02% of this energy in a direct way. Indirect factors are also important. It is estimated that up to 50% of solar energy is converted to heat and used to heat up the planet, up to 30% is used as a source of light, and the remaining portion is a source of energy that guarantees optimal material flow in the form of such fundamental processes as water, carbon, sulphur, or coal cycles (Salay 1997, Pawłowski 2011).

Solar energy may be also used to produce electricity or heat. During the year, 7,500 times more of solar energy reaches the Earth (86,000 TW) in relation to the primary energy consumed by the entire human civilization!

There are two basic methods (Kucowski, Laudyn & Przekwas 1993):

- Heliothermic method – conversion of light energy into heat used to drive the turbine and power generating unit. Solar radiation is absorbed by the solar collectors. Collectors may be either flat-plate (heating up to about 900C) or concentrating (parabolic, trough or dish – reaching 7500C in the case of dish-type collectors).
- Helioelectric method – direct conversion of solar energy into electricity. From the technical point of view, this method is based on photocells (made of silica, gallium arsenide, or cadmium sulphide).

Solar collectors should be used in areas with sunny weather throughout the year in order to produce electricity from this source.

Photocells offer more flexibility – they can convert both direct radiation and dispersed energy (i.e. they can work even on cloudy days).

In most European countries, insolation is rather moderate and subject to high variations. Therefore, solar energy is used as a secondary source of energy in water heating systems or in heating systems in buildings. Space-based solar power plants (SBSP) would solve this problem; however, the transmission of electricity back to Earth is a technical barrier.

In another interesting project, the European Union is planning on building of a 400-billion-Euro solar power plant on the Sahara desert, which could cover about 20% of the EU's energy demand (Meinhold 2009, Matlack 2010).

From the world's perspective the biggest PV plant (both in size and the amount of produced energy) was built in China and is called The Tengger Solar Park. It is located in Zhongwei, Ningxia, China. It covers 1,200 km of the Tengger desert and its output is 1,547 MW (Power Technology 2018).

Even more ambitious project is under construction in the United Arab Emirates. It is called Mohammed bin Rashid Al Maktoum Solar Park, after Vice President and Prime Minister of this country. The total capacity of the entire project is planned to reach 5,000 MW in 2030 (Power Technology 2018).

In case of the solar plants of course there is no emission of any pollution, however producing solar collectors or photocells is polluting the environment and they are hazardous waste after dismantling. This side of this technology will need further attention in the future.

5.4 *Wind power*

The first windmills were built in Persia around 2000 B.C. In Europe, they were in use since late 10th – early 11th century (Nowak & Stachel 2007). The second half of the 20th century made possible to use wind power to drive the turbines of electricity generating units.

Wind turbines are sometimes installed as standalone units to produce electricity for a single household. However, they are usually grouped in wind farms including tens or even hundreds of turbines. It means that they require large areas and by this are changing the landscape (Pawłowski 2011).

The biggest wind farm on the Earth was open in September 2018 and is called The Walney Extension. It is located between England and the Isle of Man. It has 87 turbines and may produce 659 MW of energy (Walney Extension 2019).

There are at least two basic kinds of wind turbines.

- turbines with a horizontal axis of rotation,
- turbines with a vertical axis of rotation.

One of the most important challenges connected with wind energy is the height of the turbine. Wind speed increases with the height, so the higher is the turbine the more energy may be produced. However building higher turbines is much more expensive. But there may be a solution to this which is called BAT – Altaeros Buoyan Airborne Turbine, where the turbine is installed inside of the balloon, which can fly to around 300 m above the ground level (Deodhar, Vermillion & Tkacik 2015).

Another important problem is posed by the high variability of wind conditions – both during the day/night and throughout the seasons. As a result, variations in the actual capacity of the wind farm cannot be avoided.

This problem may be eliminated by using more effective methods for the storage of electricity produced in periods when winds are stronger.

Another interesting solution is to generate hydrogen to carry energy. This ‘carrier’ could be used when winds are very weak or when there are no winds at all.

Form environmental point of view there is no emission of any pollution during the work of a turbine, however to build such turbines we need

And again, as it was in case of the solar plants, production of wind energy is not connected with any pollution of the environment, however to produce such turbines we need many rare metals like neodymium. Mining them is responsible for quite big pollution (Helder 2019), and this problem is very rarely considered.

5.5 *Geothermal energy*

Geothermal projects use the natural heat of our planet – mostly in underground waters, but also energy stored in the Earth – in underground waters, water vapour and hot rocks (Cholewa & Siuta-Olcha 2010). Liquid magma is the main source of this heat. Natural decomposition of radioactive elements is an additional factor.

The point is to create an artificial cycle: boreholes are drilled to collect hot water, this water is used, and then pumped back to the Earth for re-heating.

The use of geothermal waters, initially for therapeutic purposes, has a long history. Geothermal waters were used by the Etruscans (who built special pools in the area of present-day Italy), the Maori (present-day New Zealand), and in Ancient Rome, India, and Japan. Geothermal waters are gaining popularity as the source of energy used to produce both electricity and heat. The use of this RES depends to a large extent on the capacity of the source and temperature of water. Electricity can be produced at a temperature of 150-2000C (Pawłowski 2011). In many countries the temperature is lower, so only production of heat is possible.

The largest geothermal power producer is located in the USA – The Geysers (north of San Francisco) with 15 plants and a capacity of 1520 MW (World Atlas 2019), while Iceland (Wikipedia 2015) has the highest share of geothermal energy in a country's total energy balance – 87% of heat and 24% of electricity (near all the rest of electricity, 75,4%, is from hydropower).

5.6 *Biogas*

Biogas[1] is produced as a result of fermentation in municipal landfills, in wastewater treatment plants, or in purpose-built biogas plants (Pawłowska 1999, Bułkowska et al. 2016).

Biogas is composed mainly of methane and CO_2. Methane is particularly important, as it can produce both electricity and heat when burned (Pawłowski 2011).

Biogas burning generates a small quantity of pollution in the form of an additional emission of CO_2, but it reduces the emission of methane – a significant greenhouse gas with heat absorption coefficient much higher than CO_2.

Important advantage of using biogas is that the installation is not expensive. In the USA in 2003 the consumption of biogas from landfills was at the level of 147 trillion BTU annually (DEA 2010). In China in 2010 such consumption was 248 billion cubic meter annually (Deng et al. 2014). In Europe in 2012 the biggest production of biogas was in Germany (5067 ktoe y^{-1}), United Kingdom (1765 ktoe y^{-1}) and Italy (1096 ktoe y^{-1}).

It may be much more, especially in Eastern Europe (SEBE 2019). It is also worth to mention the foundation of European Biogas Association (EBA) in 2009 in Brussels. It is non-profit organization that is promoting biogas production and use in Europe. In 2019 they have 100 members from 28 countries (EBA 2019).

6 THE PERSPECTIVES OF SUSTAINABLE TRANSPORT

Electricity production is one of the most important problems for the energy sector, but there is another important factor – transport (Cao et al. 2016).

Around 1900 the average distance for a travel for a person per day was only 1 km. Now it is about 45 km, so 45 times more than one hundred years ago. With such growth the technology used plays important role.

Over the years, car manufacturers have ensured that their cars are becoming more and more ecological and less pollute the environment. Along with the Diesel affair which touched Volkswagen, the idea was broken. However, the conclusions have been drawn. Even in case of the new Euro 6-TEMP standard, according to which the permissible emission level of NOx is 80 mg/km, research carried out by the German ADAC (the measurement was carried out while driving) showed that many new diesel models meet these standards, among them: BMW X2 xDrive20d, Peugeot 308 SW 2.0 BlueHDi 180 (Jupp 2019).

There is a strong competition from the side of hybrid and electric cars. The most famous, and at the same time the most expensive electric cars, is Tesla 3. There is however a problem with batteries. Their capacity is shrinking over time, and they are hazardous waste after dismantling.

1. This study focuses on the energy sector. Still, certain other solutions should be mentioned. There are also biofuels, a group that also includes liquid biofuels (such as bioethanol or plant oils). Liquid biofuels may be used in cars in addition to other new solutions such as hybrid engines or fuel cells. Development of these technologies is of prime importance, as the level of emissions of pollutants from transport vehicles using fossil fuels has been increasing steadily (Tengstrom & Thynell, 1997).

In Europe in many countries there is a policy supporting such solutions (Morris 2018):

- In the Netherlands and Norway cars with combustion engines will be sold only until 2025,
- In Germany until 2030,
- In France and Great Britain until 2040.

And there are new technologies emerging. There are no problems with construction of electric cars, however there are many connected with construction of electric tracks, due to the amount of power needed. In Germany, close to Frankfurt, Siemens built electric traction, quite similar to this known from the railways. To such wires, modified trucks can connect with a pantograph (GCR 2017). Such solution is especially important in areas where we can notice huge traffic, like in Frankfurt, where cars and trucks are moving slowly, consuming more fuel, which means bigger local pollution. The length of the so called e-Highway is about 10 km, however there is going to be more. Siemens engineers say, that 40-ton truck running for 100,000 km on such electric road would save around 20 000 euro because of the reduced demand for oil.

However we must also remember of what primary energy source was used. The environmental effect of replacing diesel engine with electric one depends on how the electricity is produced. If it is produced by coal or oil it will still be polluting quite a lot.

I'm personally convinced that there is a better solution – fuel cells and using hydrogen as a fuel for transport. Such cars already exists, such as Toyota Mirai. Unfortunately political support for such technologies is extremely weak, especially if we compare it to the political support for electric cars. But it may change.

There are also technologies difficult to judge, like cars running on thorium. Such car was proposed by Laser Power Systems. Once fueled may run even 100 years without refueling (Planet 2019).

However thorium is a radioactive element, so what about the possibility of radioactive leak, for example as the consequence of car accident? So far Laser Power Systems is quite successful in case of thorium for powering industrial and residential buildings.

But the technology is not the only issue. In the sustainable development wider perspective is a must. Electric car with electricity produced by the coal burning plant will never be sustainable. But electric engine with electricity from renewables?

Good example is Rhaetian Railway, which runs in Switzerland. The most famous train in this network is Bernina Express (Rhaetian Railway 2019). It begins its journey in the city of Chur (585 m a.s.l.), than is climbing over the Bernina Pass (the highest point on the track is on the altitude of 2250 m a.s.l.) than is descending to Tirano in Italy (2,5 km from the border with Switzerland). There are 55 tunnels and 196 bridges on this route! No wonder that this railway is on the UNESCO World Heritage List. What is less known, is that the company owns not only the railway tracks but also hydropower stations that generate electricity for the trains. So it is self-sufficient, based on renewables – and by this sustainable – railway.

Another important example is hyperloop, sponsored by Elon Musk (Virgin Hyperloop 2019). Now this rather a theory, but with possibilities for becoming real in the future. Hyperloop is a train traveling with the speed around 700 mph in a tube in a vacuum. Wherever possible it is supposed to be powered by renewables, like solar installations covering the whole top of the tube.

But maybe we should thing of the transport in some other way? In may sustainable development strategies the crucial word is 'local', so local production, local materials, local sources of energy. As a result the need for transport decreases. It is known that many things may be produced everywhere. So, less transportation may appear to be much more important that new technical developments!

7 THE PERSPECTIVES OF SUSTAINABLE ENERGY IN AGRICULTURE

Agriculture plays important role in energy sector. It is not only a matter of so-called factory farming, where we have kennels of thousands of animals crammed into a small area, which –

because of the huge scale – creates need for energy. The truth is that every farm, even very small, needs energy and not only electricity. As an example, using machines, like tractors, is also connected with need for energy. It may come from fossil fuels, which means huge emissions of CO_2 and nitrogen oxides. It may also come from environmentally friendly biofuels.

Now we could observe growing need for renewable energy in agriculture. Many farmers worldwide are investing in solar energy. Many of them are also thinking about biofuels, however there are much more possibilities.

One of them is – already mentioned – biogas. In the context of agriculture local biogas plants mean (Ministry of Environment, 2019):

- Improving energy security by using renewable energy sources produced from local resources,
- The possibility of delivery of significant past of gas and electricity based on biogas,
- Production of significant amount of energy from resources not competing with the food market (which is a problem in case of energy plant crops).

Local biogas plants are almost self-sufficient and can stable deliver predictable amounts of energy.

Agriculture is also important from the perspective of climate changes. It contributes to sequestration of CO_2 by terrestrial ecosystems so is dealing with the most important greenhouse gas, emitted during combustion of coal. We made evaluation of such sequestration for Poland (Pawłowski et al. 2017). Knowing that anthropogenic CO_2 emission in Poland in 2014 was 316.8 million tons we found, that:

➡ Forests absorb 84,6 million tons, so 27% of the annual emission,
➡ Orchards 31,1 million tons, so 1%,
➡ Pastures 1,9 million tons, so 0,5%,
➡ Meadows 6,9 million tons, so 2%,
➡ Cereals 74,8 million tons, so 23,5%,
➡ Industrially cultivated plants 12,8 million tons, so 4%.

As we can see, cultivation and forests together are responsible for sequestration of 187.5 million tons of carbon dioxide, so 58% of the emission. And there are potential further opportunities to increase sequestration (including afforestation), so it can reach even 80% of country's CO_2 emission. So, agriculture may play important role in mitigation of climate changes connected with CO_2 emissions, especially in countries which are still producing energy from coal.

The new challenge is connected with an interesting phenomenon: rural areas are increasingly becoming multifunctional, where traditional agriculture is only one of many development opportunities (complemented by service and industrial sectors). Villages more and more often serve also as new estates for people working in cities, but wanted to live outside of them (Pawłowski, 20110. This means that in the future the structure of energy use will be the same, both in case of cities and villages. To put it other way, the total demand for energy will grow.

8 CONCLUSIONS

Strong support for renewables means that probably sooner or later they will become the main source of energy on the Earth. From the perspective of sustainable development not only ensuring security for the growing demand for energy is a challenge. There is also a tricky issue of energy efficiency, even on the level of households. In the past we had no energy saving devices, however we were using very few devices at all. Now we have many energy saving devices, however we are using so many of them, that the total energy demand from our households is growing! And it is also true, that production of energy saving device sometimes requires more energy than this device will earn during its lifetime! We should also remember about so-called Jevons Paradox, according to which a complete confusion of concepts is that the effective use of fuels is equivalent to reduction of their consumption (Pieńkowski 2012). Completely opposite dependence is true. That was in the past and it will be in the future. So, this paradox occurs, when technological progress increases the efficiency with which a resource is used (reducing the amount necessary for any use), but the rate of consumption of that resource rises due to increasing demand. Jevons was writing about coal, however in our modern world we may say the same about the demand for electricity. Strategies supporting renewables should also put more pressure on education how to use less energy.

It is also possible that a radical discovery of a new energy source will change the scenario. The energy of the Universe generated as a result of the Big Bang is one theoretical option (Michnowski 2007). Is it possible that one day we will be able to capture and use it to the benefit of the entire human race in a sustainable way?.

REFERENCES

Borkowski, R. 1991. Przed Czarnobylem był Kysztym. Aura 4: 30.

Borys, T. 2005. Wskaźniki zrównoważonego rozwoju. Warsaw, Białystok, EiS.

Boyle, G. 1996. Renewable Energy: Power for Sustainable Future. Oxford: The Open University and Oxford University Press.

Brueck, H. 2018. H. The world's largest wind farm was just completed in the Irish Sea – and it's more than twice the size of Manhattan. Business Insider, September 13.

Bułkowska, K., Gusiatyn, Z.M., Klimiuk, E., Pawłowski, A., Pokój, T. 2016. Biomass for Biofuels. Boca Raton, London, New York, Leiden: CRC Press, Taylor & Francis Group, A Balkema Book.

Cao, Y., Pawłowski, A. 2013. Biomass as an Answer to Sustainable Energy: opportunity versus Challenge, in Environment Protection Engineering. *Environment Protection Engineering* 39 (1): 153–161.

Cao, Y., You, I., Shi, Y., Hu, W., 2016. Evaluation of Automobile Manufacturing Enterprise Competitiveness from Social Responsibility Perspective. *Problemy Ekorozwoju/Problems of Sustainable Development* 11(2): 89–98.

CEF (Conserve Energy Future), 2019. Is renewable energy sustainable?, https://www.conserve-energy-future.com/IsRenewableEnergySustainable.php [1.06.2019].

DEA, 2010. Alternative Fuels Data Center, What is biogas? https://afdc.energy.gov/fuels/natural_gas_re newable.html [1.06.2019].

Deng, Y., Xu, J., Liu, Y., Mancl, K., 2014. Biogas as a sustainable energy source in China: Regional development strategy application and decision making. *Renewable and Sustainable Energy Reviews 35*: 294–303.

Deodhar, N., Vermillion, C., Tkacik P. 2015. A case study in experimentally-infused plant and controller optimization for airborne wind energy systems. IEEE Xplore: July 30.

Drastichová, M., 2017. Decomposition Analysis of the Greenhouse Gas Emissions in the European Union. *Problemy Ekorozwoju/Problems of Sustainable Development* 12(2): 27–35.

EBA, 2019. European Biogas Association. http://european-biogas.eu/[1.06.2019].

E&C Consultants, 2019. Making the right choices in sustainable energy management, https://www.eecc.eu/sustainable-energy-management [1.06.2019].

European Comission, 2009. 2020 climate & energy package. Brussels.

European Comission, 2018. 2030 climate & energy framework. Brussels.

Edwards, R. 2002. Japan's Nuclear Safety Dangerously Weak. New Scientist Tech 1 October.

EPA, 2019. Overview of Greenhouse Gases, https://www.epa.gov/ghgemissions/overview-greenhouse-gases [1.06.2019].

GCR, Global Construction Review, http://www.globalconstructionreview.com/news/germany -build-10-km-electric-highway/[16.08.2017]

Global Energy Statistical Yearbook, 2018, https://yearbook.enerdata.net/[1/06/2019].

Gore, A., 2017. An Inconvenient Sequel. Rodale Books.

Harris, J.M., Wise, T.A., Gallagher, K.P. & Goodwin, N.R. (eds). 2001. A Survey of Sustainable Development. Social and Economic Dimensions. Washington, Covelo, London: Island Press.

International Energy Annual, 2005, 2015, Washington D.C.

Johansson, T., Pirouzfar, P. 2019. Sustainability Challenges in Energy Use Behavior in Households: Comparative Review of Selected Survey-based Publications from Developed and Developing Countries. *Problemy Ekorozwoju/Problems of Sustainable Development* 14(2):33–44.

Helder, M., 2019. Renewable energy is not enough: it needs to be sustainable. World Energy Forum. https://www.weforum.org/agenda/2015/09/renewable-energy-is-not-enough-it-needs-to-be-sustainable/ [1.06.2019].

IRENA (International Renewable Energy Agency, 2018, Renewable energy highlights, https://irena.org/media/Files/IRENA/Agency/Publication/2018/Jul/IRENA_Renewable_energy_ highlights _July_2018.pdf?la=en&hash=F0E22210DEB43512673D6A573C1879F10CFC41D0 [1.06.2019].
Ivanow, I. 2017. Sugarcane Cultivation in Brazil: Challenges and Opportunities, https://medium.com/remote-sensing-in-agriculture/sugarcane-cultivation-in-brazil-challenges-and-opportunities-fd93ca037e8d [1.06.2019].
Jackson, J.P. 1996. Unending Nightmare. Time 6 May: 23-25.
Jaśkowski, J. 1988. Luki i niedomówienia. Nowy Medyk 16 December: 8-9.
Jupp. E, 2019, New diesel cars that emit almost ZERO NOx. Motoring Research, https://www.motoringresearch.com/car-news/diesel-cars-zero-nox/[1.06.2019].
Kalinowska, A. 1992. Ekologia – wybór przyszłości. Warsaw: Editions Spotkania.
Kucowski, J., Laudyn, D. & Przekwas, M. 1993. Energetyka a ochrona środowiska. Warsaw: WNT.
Kumar, P., Sharma, H., Pal, N., Sadhu, P.K, 2019, Comparative Assessment and Obstacles in the Advancement of Renewable Energy in India and China. *Problemy Ekorozwoju/Problems of Sustainable Development* 14(2):191–200.
Lindzen, R.S., 2010. Global Warming: The Origin and Nature of the Alleged Scientific Consensus. *Problemy Ekorozwoju/Problems of Sustainable Development* 5(2): 13–28.
Maćkow, J., Paczosa, A. & Skirmuntt, G. 2004. Eko-generacja przyszłości. Katowice, Warsaw: WNS.
Matlack, C. 2011. A Consortium Wants to Invest $ 560 Billion in Sahara Solar Panels. http://www. businessweek.com/magazine/content/10_38/b4195012469892.htm [1.03.2011].
Meinhold, B. 2011. World's Largest Solar Project Planned for Saharan Desert. http://inhabitat.com/worlds-largest-solar-project-sahara-desert/[1.03.2011].
Michnowski, L. 2007. Społeczeństwo przyszłości a trwały rozwój. Warsaw: KPP2000P PAN.
Ministry of Environment, 2019. Polityka energetyczna Polski do 2030 r., Warsaw.
Moberg, A. 1986. Before and After Chernobyl. Nuclear Power in Crisis. A Country by Country Report. Malmo: Team Offset.
Morris C., 2018, 8 European Countries & Their EV Policies. Clean Technica, https://cleantechnica.com/2018/11/04/8-european-countries-their-ev-policies/[1.06.2019].
New Details on Japan Nuclear Accident. 1999. Science Daily 6 December.
Nowak, W. & Stachel, A.A. 2007. Ocena możliwości korzystania z energii wiatru w Polsce na tle krajów Europy i Świata, http://www.fundacjarozwoju.szczecin.pl/biuro/teksty2/FRPZ%20-%20Referat%20Nowak+Stachel.pdf [30.01.2011].
O'Neill, P. 1997. *Environmental Chemistry*. New York: Chapman and Hall.
Pawłowska, M. 1999. Możliwość zmniejszenia emisji metanu z wysypisk na drodze jego biochemicznego utleniania w rekultywacyjnym nadkładzie glebowym – badania modelowe. Lublin: Politechnika Lubelska.
Pawłowski, A., 2011. Sustainable Development as a Civilizational Revolution. Multidimensional Approach to the Challenges of the 21st century, Boca Raton, Londyn, Nowy Jork, Leiden: CRC Press, Taylor & Francis Group.
Pawłowski A., Pawłowska M., Pawłowski L., 2017. Mitigation of Greenhouse Gases Emissions by Management of Terrestrial Ecosystem. *Ecological Chemistry and Engineering. S = Chemia i Inżynieria Ekologiczna*. S 24(2): 213–222.
Pawłowski A., 2010. Sustainable Development vs Environmental Engineering: Energy Issues, in: Nathwani J., Ng A. (ed.) Paths to Sustainable Energy, InTech, Rijeka: 13–28.
Phelps, J., Blackford, J., Holt, J., Polton, J. 2015. Modelling Large-Scale CO_2 Leakages in the North Sea, *International Journal of Greenhouse Gas Control*, 38: 210–220.
Pieńkowski, D., 2012, The Jevons Effect and the Consumption of Energy in the European Union. *Problemy Ekorozwoju/Problems of Sustainable Development* 7(1): 05–116.
Piłat, B., 2015. Blue coal, czyli ekologiczny węgiel po polsku. Gazeta Wyborcza April 7.
Planet, C., 2019. Car Runs 1 Million Miles on 8 Grams Of Thorium, https://www.captain-planet.net/thorium-friend-or-faux-car-runs-1-million-miles-on-8-grams/[1.06.2019].
Power Technology, 2018, The world's biggest solar power plants, https://www.power-technology.com/features/the-worlds-biggest-solar-power-plants/[1.06.2019].
Rhaetian Railway, 2019. https://www.rhb.ch/en/panoramic-trains/bernina-express [1.06.2019]
Salay, J. 1997. Energy, From Fossil Fuels to Sustainable Energy Resources. Uppsala: The Baltic University Programme.
SEBE, 2019. Sustainable and Innovative European Biogas Environment. https://arquivo.pt/wayback/20141128141853/http://www.sebe2013.eu/home/about [1.06.2019].
Singh, M. 2019. Social Justice and Sustainable Development. *Problemy Ekorozwoju/Problems of Sustainable Development* 14(2):57–62.

Sztumski, W. 2019. For Further Social Development, Peaceful, Safe and Useful for People. *Problemy Ekorozwoju/Problems of Sustainable Development* 14(2):25–32.
Tengstrom, E. & Thynell, M. 1997. Towards Sustainable Mobility. Transporting People and Goods in the Baltic Region. Uppsala: Ditt Tryckeri.
The Need For Change. The Legacy of TMI: Report of the President's Commission on the Accident at Three Miles Island. 1979. The Commission on the Accident: Washington.
United Nations, 2015. Sustainable Development Goals, New York.
United Nations Global Compact, 2012. Sustainable Energy for All: Opportunities for the Utility Industry. Accenture, New York.
Universal Encyclopedia. 1984, 2000. PWN, Warsaw.
Velkin, V.I., Shcheklein, S.E., 2017. Influence of RES Integrated Systems on Energy Supply Improvements and Risks. *Problemy Ekorozwoju/Problems of Sustainable Development* 12(1): 123–129.
Vince, G. 2002. Safety Violations Shut Dutch Nuclear Reactor. New Scientist Tech February 4.
Virgin Hyperloop One, https://hyperloop-one.com/[1.06.2019].
Vlosky, R., Smithhart, R., 2011. A Biref Global Perspective on Biomass for Bioenergy and Biofuels. *Journal of Tropical Forestry and Environment* 1(1): 1–13.
Walney Extension, 2019. https://walneyextension.co.uk/[1.06.2019].
WCED. 1987. Our Common Future. The Report of the World Commission on Environment and Development. New York: Oxford University Press.
Wikipedia. http://ww.wikipedia.org [15.03.2011].
World Atlas, 2019, Largest geothermal power plants in the World, https://www.worldatlas.com/articles/largest-geothermal-power-plants-in-the-world.html [1.06.2019].
Żelazna, A., Gołębiewska, J., 2015. The Measures of Sustainable Development – a Study Based on the European Monitoring of Energy-related Indicators. *Problemy Ekorozwoju/Problems of Sustainable Development* 10(2): 169–177.

The Role of Agriculture in Climate Change Mitigation – Pawłowski, Litwińczuk & Zhou (eds)
 ISBN 978-0-367-43372-7

Cultivation of catch crop as a sustainable way of reducing CO_2 content in the atmosphere

L. Pawłowski, M. Pawłowska, A. Pawłowski, W. Cel, K.Wójcik Oliveira & M. Piątek
Faculty of Environmental Engineering, Lublin University of Technology, Lublin, Poland

C. Kwiatkowski & E. Harasim
University of Life Science in Lublin, Poland

G. Zhou & L. Wang
Zhejiang A and F University, Zhejiang, China

1 INTRODUCTION

The average global temperature is still growing (IPCC 2013), which is correlated with increasing concentration of green-house gases (GHG), mainly CO_2 in the atmosphere. Various actions are taken for mitigation of GHG emission. But, taking into account the current trends in socio-economic development, reduction of CO_2 emission will not be easy (Xu & Li 2018). Development of our civilization is inseparably connected with growing demand of energy while the energy production is still basing on fossil fuels (Hui-Ming et al. 2012, Johansson & Pirouzfar 2019). Combustion of these fuels is the main anthropogenic source of the emission of CO_2 (IPCC 2013). Therefore, the gradual insertion of renewable sources to the energy production systems should be implemented (Lata-Garcia et al. 2018). At the same time, it is necessary to seek more efficient CO_2 sinks.

The control of CO_2 exchange between the atmosphere and terrestrial and marine ecosystems is promising way for removing CO_2 from the atmosphere. It was observed that the net amount of the CO_2 absorbed in ocean water increased from 3.7 Gt CO_2/yr in 1960-1969 to 9.5 Gt CO_2/yr in 2016 (Trujillo & Thurman 2017). However, this sink will decrease with time, as a result of ocean water acidification. Within the period from 1700 to 1994, the pH of ocean water dropped from 8.25 to 8.14 (Jacobson, 2005). A decrease in the pH of ocean water negatively influences the ocean ecosystem, among others it cause the damage of the coral reefs. Attempts are made to mitigate this phenomena by enhancing the absorption of CO_2 by algae in the photosynthesis process. Fertilization of ocean water with Fe ions are carried out in order to achieved this purpose. Intensification of algae growth enhances the absorption of CO_2 by ocean water and has a beneficial effect on fishing (Cai 2018).

The studies carried out by Le Quére et al. (2018) shows that starting from the 1960s, the share of another important sink, i.e. net absorption of CO_2 by plants is growing from 5.1 Gt CO_2/yr in 1960-69 to 11.0 Gt CO_2/yr in 2007-2016, reaching 8.9 Gt CO_2/yr in 1970-1979, 9.2 Gt CO_2/yr in 1990-1999 and 10.6 GtCO_2/yr in 2000-2009. In 2016, its value reached at 9.9 GtCO_2/yr, which was lower than the average from the previous decade.

There are further possibilities of reducing atmospheric CO_2 concentration with using terrestrial ecosystems, such as the decrease of CO_2 emission from soils, by slowing down the oxidation of soil organic compounds, e.g. by applying the crop rotation, cover cropping, catch cropping, intercropping, mulching, and no tillage practice (Wang et al, 2010) or the increase of CO_2 absorption by plants in the photosynthesis process.

1.1 *Characteristics of catch crops*

Enhancement of carbon sequestration is one of the main tasks of a sustainable agriculture. This agricultural system bases on the soil conservation practices, such as: limiting the soil disturbance (no-till farming) or permanent soil covering by plant biomass combined with the crops rotations (Hobbs et al. 2007).

The catch crops and intercrops are used to prevent the wind and water erosion (Hobbs et al. 2007), thus the name "cover crops" is also suitable for them.

Catch crops are the fast-growing plant species that are grown between the successive main crops (Lockhart & Wiseman 2014), and intercrops are the two or more crops cultivated in the same space at the same time (Lithourgidis et al. 2011). The name "catch crops" refers to the possibility of the plants to take up and retain the nutrients from the soil, that prevents leaching out these substances from the surface to deeper layers.

Catch crops can be cultivated in pure or mixed sowing. There are mainly the plants with a short vegetation period. The legumes (e.g. clover, pea and vetch), grasses (e.g. rye, ryegrass, wheats, barley), and mustards are usually cultivated as the catch crops in the temperate region (Wang et al. 2010).

The appropriate selection of plant species used as the catch crops is necessary to achieve the sufficient yield under the specified soil properties and environmental conditions. The species used in the mixture should have different types of root system, which are characterized by different depth. This causes that the soil is enriched with organic substances on several levels. Additionally, such a differentiation allows the nutrients from different depths to move up the soil profile, along the roots, to the above-ground parts, where they become again available for the main crop plants (Thorup-Kristensen 2006). An important function of catch crops is improve the soil physical properties, such as aggregation, porosity, water and air conditions, due to increase of humus content. Impact of cover crops on soil water content is explicit. During the vegetation season the cover crops decrease soil water content, because of transpiration, but after ploughed or using as mulch they increase the soil water retention (Qi & Helmers 2010).

Depending on the sowing time, three types of catch crops can be distinguished:

- stubble crops – seeds are sown in summer, while the crops are harvested in autumn for green forage, mowed and filched or plowed without mowing. After mowing, plants may be also left for winter in the form of mulch. The plants which are most commonly cultivated as this type of intercrops are: brassicas (white mustard, black mustard, rapeseed, oilseed radish, stubble turnip, *Brassica oleracea var. medullosa*), legumes (horse bean, yellow lupin, narrowleaf lupin, sugarsnap peas, field peas, spring vetch, serradella), and other species (blue tansy, sunflower, oat)
- undersown crops – spring cereals sown in spring or, rarely sown with winter cereals
- winter catch crops – sown at the autumn and harvested in spring of the next year. According to their harvest period the plants are divided on: early (winter agrimony, winter rapeseed), intermediate (winter rye), and late (*Lolium westerwoldicum*, crimson clover) (Finch et al. 2014).

The appropriate selection of the plant species allows for achieving a high yield of plant biomass under certain field conditions. The yields of shoot and root biomass of the selected catch crops are presented in Table 1. The results showed a significant differentiation of the yields, both between the plant species, and also between the same species under different locations.

1.2 *The role of catch crops in carbon sequestration*

Use of the ground-cover plants is a promising option for sequestering the carbon in the agricultural systems (Wang et al., 2010; Popelau et al.2015). This practice allows to increase the nutrients content in the soil that improves the productivity of the main crop (Wang et al. 2010), leading to increase carbon binding in the biomass. Potential of catch crops for carbon sequestration also refers to the retention of soil organic carbon (SOC) in the upper layer of the soil. Sainju et al. (2006) observed that SOC content in the layer 0 to 10 cm of the soil covered by vetch and rye crops increased by 6-8% within 3 years. In this time, SOC content in this layer in no-tilled plots decreased by 5%. Popelau et al. (2015) analyzed the effect of cover crops on soil carbon content on the basis of the literature data. According to their estimation the global potential of carbon sequestration due to catch crops is 0.12 Pg C yr^{-1}. They stated that soil saturation with C will occur after 155 years of cultivation. Therefore, such a practice can be long-term sink for the atmospheric carbon.

In the cover cropping system the plant biomass is left in the field as a mulch or green manure. This biomass becomes the source of humic substance and nutrients. Tang et al. (2017) showed that winter cover crop residues caused the significant increase of carbon content in the humins (C-HUM), humic acid (C-HAF) and fulvic acid (C-FAF) in the paddy soil.

The above-ground biomass of catch crops can be also used as a substrate for biofuels production, e.g. a biogas in the anaerobic digestion process or syngas, bio-oil and biochar in the pyrolysis process. This type of biomass management is not a sequestration process, but it helps to reduce CO_2 emissions by using a renewable carbon source. Thus, it also mitigate the green-house effect.

Some amount of catch crops biomass (mainly the below-ground ones) remains in the soil even than the shoots are taken away from the field. It becomes a source for humus. The root biomass is usually lower than the shoot biomass (Table 1), but in the case of some plants the insignificant differences have been observed. For example, the mass of roots of perennial

Table 1. Yield of selected plant species used as catch or cover crops.

Plant species	Yield Above-ground biomass [t d.m. ha^{-1}]	 Below-ground biomass [t d.m. ha^{-1}]	Sowing time/ location	Reference
Hairy vetch (*Vicia villosa* Roth.) and rye (*Secale cereale* L.), mix	5.6 to 8.2	0.32 to 0.88	Autumn/ USA Georgia	Sainju et al. (2006)
Hairy vetch (*Vicia villosa* Roth.)	2.4 - 5.2	0.047 - 0.656		
Rye (*Secale cereale* L.)	2.3 - 6.1	0.174 - 0.772		
Ryegrass (*Lolium multiflorum* Lam.)	6.66 - 9.37	-	Autumn/ Turkey Samsun	Yavuz et al. (2017)
Fodder radish (*Raphanus sativus*)	1.7	1.3	Autumn/ Denmark Foulum	Li et al. (2015)
Winter vetch (*Vicia villosa* Roth.)	1.7	1.2		
Red clover (*Trifolium pretense* L.)	1.9	1.4	Spring/ Denmark Foulum	
Perennial ryegrass (*Lolium perenne* L.)	1.3	1.3		
Ryegrass (*Lolium multiflorum* Lam.) and red clover (*Trifolium pretense* L.), mix	2.1	1.5		
Winter rye (*Secale cereale* L.) and hairy vetch (*Vicia villosa Roth.*) mix	3.3	1.4	Atumn/ Denmark Aarslev	Chirinda et al. (2012)
Phacelia (*Phacelia tenacetifolia* Benth.)	4.2	0.5	Autumn/ Denmark Aarslev	Thorup-Kristensen (2001)
Oats (*Avena sativa* L.)	3.1	0.7		
Winter rape (*Brassica napus* L.)	4.0	1.4		
Fodder radish (*Raphanus sativus*)	4.7	0.9		
Black oats (*Avena strigose*)	7.98	1.85	Winter/ Rio Grande do Sul Brazil	Redin et al. (2018)
Ryegrass (*Lolium multiflorum*)	5.48	1.15		
Vetch (*Vicia sativa*)	3.67	1.33		
Pea (*Pisum arvensis*)	5.43	0.66		
Blue lupine (*Lupinus angustifolius*)	5.54	0.85		
Native lupine (*Lupinus albescens*)	5.51	0.76		
Oilseed radish (*Raphanus sativus oleiferus*)	6.81	0.86		

ryegrass (*Lolium perenne* L.) sown in spring season in Denmark was 50% of the total dry mass. The mass ratio of below-ground parts to total biomass for fodder radish (*Raphanus sativus*), vetch (*Vicia villosa* Roth.), red clover (*Trifolium pretense* L.) cultivated in this same location was ca. 40% (Li et al. 2015). The significantly lower values of the ratio: 10.6 and 18.4%, respectively, were observed in the cases of phacelia (*Phacelia tenacetifolia* Benth.) and oats (*Avena sativa* L.). These latter values are typical for the most plant species. According to Fageria (2013) root biomass normally contribute from 10% to 20% of total plant mass.

The yield of root biomass depends not only on the species diversity. It is also affected by land management system, which determines the soil properties. Hu et al. (2018) observed that the average root biomass of different species of catch crops cultivated in Denmark was significantly higher in organic farming system than in conventional farming system. These value were on average 1.27 t d.m ha^{-1} and 0.75 t d.m. ha^{-1}, respectively. The average yields of root biomass of the winter and summer cover crops examined in Brasil was 1.07 and 0.95 t d.m ha^{-1}, respectively (Redin et al. 2018). Additionally, the root and shoot biomass was similar in order to carbon content (Table 2), although the different species were planted in each sowing season.

Considering the few times higher (averagely 4-times) yield of above-ground biomass in comparison to the below-ground part (Table 1), the shoot biomass is significantly larger sink of carbon.

1.3 *Potential of carbon sequestration by catch crops – The Polish case study*

The average yield of catch crops cultivated in Lublin region (eastern Poland) is 3.4 t d.m ha^{-1} (Table 3). It is less than the mean value determined by Redin et al. (2018) for winter cover crops cultivated in Brasil, that was 5.8 t d.m ha^{-1}, but greater than the mean value achieved by Li et al. (2015) in Denmark for winter and spring sown crops (Table 1).

According to data of Central Statistical Office of Poland (2019) approximately 7.8 mln hectares of cereals were cultivated in Poland in 2018. This area is potentially suitable for catch crops cultivation. Taking into account the average yields of biomass (total dry mass of shoots and carbon content) of the selected plants cultivated as catch crops in Poland (Table 3) it was estimated that if the catch crops were cultivated over the entire sown area, it would be possible to achieve 20.6 - 32.2 mln t of above-ground biomass within the year, depending on the plant species. These plants will be able to absorb from the atmosphere 8.7 - 14.1 mln t C/yr in shoot biomass. But, assuming that the root biomass is about 10% of the total plant dry mass (the lower value given by Fageria 2013) and the data on the root yield given in Table 1, we calculated that 2.3 mln - 3.6 mln t of root-derived organic matters would be enter to the soil each year. This biomass will undergo the humification, leading to production of the stable organic substances.

The humification coefficient (parameter that refers to the part of organic matter that is converted to the humic substance) for roots is estimated on 0.42 (Bonten et al. 2014), thus 1.0 - 1.5 mln t of initial root-derived biomass will remain in the soil. It corresponds to 0.5 - 0.7 mln t of

Table 2. Carbon content in shoot and root biomass of catch crops from the winter sowing (Redin et al. 2018).

Species	Carbon content [% d.m.]		
	Shoot biomass (C_S)	Root biomass (C_R)	C_R/C_S ratio
Black oats (*Avena strigose*)	47.8	46.1	1.04
Ryegrass (*Lolium multiflorum*)	46.4	45.8	1.01
Vetch (*Vicia sativa*)	48.0	46.9	1.02
Pea (*Pisum arvensis*)	47.0	46.4	1.01
Blue lupine (*Lupinus angustifolius*)	47.6	46.0	1.03
Native lupine (*Lupinus albescens*)	47.3	46.7	1.01
Oilseed radish (*Raphanus sativus oleiferus*)	45.9	46.2	0.99
Average value ± SD	*47.1±0.8*	*46.3±0.4*	*1.02*

Table 3. Quantitative characteristics of catch crops cultivated in Poland (Pawłowska et al. in press).

Species	Biomass yield [t d.m./ha]	Carbon content [% d.m.]	Carbon yield [t C/ha]
White mustard (*Sinapis alba*)	4.26	42.5	1.81
Winter rye (*Secale cereale*)	4.07	42.5	1.73
Blue tansy (*Tanacetum annuum*)	3.98	42.7	1.70
Spring vetch + field pea (*Pisum sativum*)	3.46	42.5	1.47
Spring rapeseed (*Brassica napus*)	3.40	42.6	1.45
Oats (*Avena sativa*) + spring vetch + field pea *(Pisum sativum)*	3.22	42.5	1.37
Serradella (*Ornithopus sativus*)	3.05	42.6	1.30
Red clover (*Trifolium pretense*)	2.69	42.8	1.15
Narrowleaf lupin (*Lupinus angusti-folius*) + field pea *(Pisum sativum)*	2.64	42.4	1.12
Average value	*3.4*	*42.6*	*1.5*

carbon (assuming the average C content in root biomass of 46.3% - Table 2), that would be sequestrated in the Polish soils each year.

The above-ground biomass of catch crops is usually used as green fertilizer for improving soil quality. Assuming that entire stock of annual shoot biomass of catch crops would be leave in the fields, and a humification coefficient for above-ground crop residues is 0.22 (Bonten et al. 2014), we estimated that 4.5 - 7.1 mln t of shoot-derived biomass will remain in the soil. It corresponds to 1.9 – 3.0 mln t of carbon (assuming the average C content in root biomass of 42.6% - Table 2), that would be sequestrated in the Polish soils each year.

Summarizing, annually 2.4 - 3.7 million t of carbon could be permanently deposited in soils under the cereal cropping in Poland due to common using of catch crops. It gives an annual increase in the carbon content in these soil by 0.01%. Assuming that the average carbon content in Polish soils is 1.12% (IUNG, data from 2015) and the target value is 2.31% (which corresponds to 4% of organic matter content, when C concentration in humus is 57.7%) the time needed to achieve this goal is 119 years.

2 CONCLUSION

The catch crop cultivation seems to be a viable way for reducing CO_2 concentration in the atmosphere. Popularization of this commonly known agricultural practices does not involve the new technologies, but it would supply the multifaceted benefits for the environment. It enables to:

- increase the carbon sink, by carbon retention in the catch crops biomass and long-term carbon sequestration in the soil humus,
- improve the soil fertility, that leads to increase of carbon sequestration in a biomass of the main crops,
- prevent the CO_2 emissions from the burning of fossil fuels due to their partial replacement by the biomass of catch crops, used as an alternative energy source.

REFERENCES

Bonten L.T.C., Elferink E.V., Zwart K. BioESoil, version 0.1. *Tool to assess effects of bio-energy on nutrient losses and soil organic matter.* Alterra Wageningen UR.

Cai H. 2018. Algae-Based carbon sequestration. *Earth and Environmental Science.* 120(012011).

Chirinda N., Olesen J.E., Porter J.R., 2012. Root carbon input in organic and inorganic fertilizer-based systems. *Plant Soil.* 359, 321–333.

Fageria N.K. *The Role of Plant Roots in Crop Production.* CRC Press Taylor & Francis. Boca Raton, FL. 2013.

Finch H.J.S., Samuel A.M., Lane G.P.F. *Lockhart & Wiseman's Crop Husbandry Including Grassland.* 9th Edition. Elsevier Ltd. 2014. ISBN 978-1-78242-371-3.

Hobbs P.R., Sayre K., Gupta R. 2007. The role of conservation agriculture in sustainable agriculture. *Philos. T. Roy. Soc. B.* 363 (1491).

Hu T., Sørensen P., Wahlström M.E., Chirinda N., Sharif B., Li X., Olesen J.E. 2018. Root biomass in cereals, catch crops and weeds can be reliably estimated without considering aboveground biomass. *Agr. Ecosyst. Environ.* 251:141–148.

Hui-Ming W., Wen-Hsiung Y., Chao-Wu C., Padilan M.V. 2012. Renewable energy supply chains, performance, application barriers, and strategies for further development. *Renewable and Sustainable Energy Reviews.* 16(8):5451–5465.

IPCC. *Fifth Report, Climate Change* 2013: The physical science basis. https://www.ipcc.ch/report/ar5/wg1/.

IUNG, Monitoring chemizmu gleb w Polsce, https://www.gios.gov.pl/chemizm_gleb/index.php?mod=wyniki&cz=CCh,G, 201.

Jacobson M.Z. 2005. Studying ocean acidification with conservative, stable numerical schemes for nonequilibrium air-ocean exchange and ocean equilibrium chemistry. *J. Geophys Res-Atmos.* 110 (D7).

Johansson, T., Pirouzfar, P. 2019. Sustainable challenges in energy use behavior in households: comparative review of selected survey-based. Publications from developed and developing countries. *Probl. Ekorozw ./Problems of Sustainable Development* 14(2):33–44.

Lata-Garcia, J., Reyes-Lopez, C., Jurado, F., 2018. Attaining the Energy sustainability: analysis of the Ecuadorian strategy. *Probl. Ekorozw./Problems of Sustainable Development.* 13(1):21–29.

Le Quere et al. 2018. Global Carbon Budgets 2017. *Earth Syst. Sci. Data.* 10:405–448.

Li, X., Petersen, S.O., Sørensen, P., Olesen, J.E. 2015. Effects of contrasting catch crops on nitrogen availability and nitrous oxide emissions in an organic cropping system. *Agric. Ecosyst. Environ.* 199, 382–393.

Lithourgidis A.S., Dordas C.A., Damalas C.A., Vlachostergios D.N. 2011. Annual Intercrops: An Alternative Pathway for Sustainable Agriculture, *Aust. J. Crop Sci.* 5(4):396–410.

Lockhart J.A.R., Wiseman A.J.L., *Introduction to Crop Husbandry.* Pergamon Press. 2014.

Pawłowska M, Pawłowski A., Pawłowski L., Cel W. Wójcik-Oliveira K., Kwiatkowski C. et al. Possibility of carbon dioxide sequestration by intercrops. *Ecol. Chem. Eng.* S, in press.

Poeplau C., Don A., Carbon sequestration in agricultural soils via cultivation of cover crops – A meta-analysis. 2015. *Agriculture, Ecosystems & Environment*, 200:33–41.

Qi, Z., and M.J. Helmers. 2010. Soil water dynamics under winter rye cover crop in central Iowa. *Vadose Zone* J. 9:53–60.

Redin M., Recous, S. Aita C., Chaves B., Pfeifer I.C., Bastos L.M., Pilecco, G.E., Giacomini S.J. 2018. Root and Shoot Contribution to Carbon and Nitrogen Inputs in the Topsoil Layer in No-Tillage Crop Systems under Subtropical Conditions. *Rev. Bras. Ciênc. Solo* 42.

Sainju U.M., Singh, B. P. Wayne F. W., Wang S. 2006. Carbon Supply and Storage in Tilled and Nontilled Soils as Influenced by Cover Crops and Nitrogen Fertilization. *J. Environ. Qual.* 35(4):1507–17.

Statistical Data. *Land use and sown area*, Statistics Poland, Agriculture Department. Warsaw 2019 (https://stat.gov.pl › uzytkowanie_gruntow_i_powierzchnia_zasiewow_w_2018).

Tang H. Xiao X., Tang W., Wang K. Li C., Cheng K. 2017. Returning winter cover crop residue influences soil aggregation and humic substances under double-cropped rice fields, *Rev. Bras. Ciênc. Solo* 41.

Thorup-Kristensen K. 2006. Effect of deep and shallow root systems on the dynamics of soil inorganic N during 3-year crop rotations. *Plant Soil* 288(1):233–248.

Thorup-Kristensen, K., 2001. Are differences in root growth of nitrogen catch crops important for their ability to reduce soil nitrate-N content, and how can this be measured? *Plant Soil* 203:185–195.

Trujillo A.P. & Thurman H. *Essentials of Oceanography.* 12th Edition. Pearson 2017.

Wang Q., Li Y., Alva A., 2010. Cropping systems to improve carbon sequestration for mitigation of climate change. *Journal of Environmental Protection*, 2010, 1, 207–215.

Xu J. & Li J. 2018. The tradeoff between growth and environment: evidence from China and the United States. *Probl. Ekorozw./Problems of Sustainable Development* 13(1):15–20.

Yavuz T., Sürmen M., Albayrak S., Çankaya N. 2017. Determination of Forage Yield and Quality Characteristics of Annual Ryegrass (*Lolium multiflorum* Lam.) Lines, *Tarım Bilim Derg.* 23(2):234–241.

The Role of Agriculture in Climate Change Mitigation – Pawłowski, Litwińczuk & Zhou (eds)
 ISBN 978-0-367-43372-7

Biochar for greenhouse gas mitigation

W. Yuan
Department of Biological and Agricultural Engineering, North Carolina State University, Raleigh, USA

H. Zhang
College of Environment and Resources, Zhejiang Agricultural and Forestry University, Hangzhou, China

X. Jia
The Graduate School, Chinese Academy of Agricultural Sciences, Beijing, China

A. James
Department of Mechanical Engineering, Universidad Tecnológica de Panamá, El Dorado, Panamá

M. Wang
The Institute of Agricultural Resources and Regional Planning, Chinese Academy of Agricultural Sciences, Beijing, China

1 INTRODUCTION

Biomass gasification is considered one of the most promising technologies for production of high-quality biofuels. It is a process in which biomass undergoes incomplete combustion to produce a gas product called syngas (or producer gas) that consists mainly of H_2, CO, CH_4, CO_2, and N_2 (if air is used in combustion) in various proportions. Biomass gasification has many advantages over direct combustion. It converts low-value feedstocks to high-quality combustible gases, which can be not only directly burned or used for heat or electricity generation but also turned into liquid transportation fuels, such as Fischer-Tropsch (FT) diesel and gasoline (Dunn et al. 2004, Piemental 2012, Dowbor 2013, Kumur et al. 2019). The FT process has been successfully employed at large scales; for example, it was used in Germany during WWII, and it is used in South Africa to produce oil and gasoline from coal and in commercial plants by Shell Oil to produce diesel fuel from natural gas. The FT process produces high-value, colorless, odorless, and low-toxicity clean-burning fuels that can be used in conventional engines with little or no modification.

In addition to syngas, gasification also generates biochar, but typical biochar yields are low, e.g., only ~5% of the feedstock (Demirbas 2004). As a variation of normal gasification, top-lit updraft (TLUD) gasification has the remarkable ability to simultaneously produce biochar and syngas. Several researchers have conducted studies of various designs of top-lit updraft gasifiers (Tryner et al. 2014, Birzer et al. 2013). However, all previous work lacks a full prospective of the process from the perspective of product quantification and process optimization. Little is known about this thermochemical process. Field studies have shown mixed effects of biochar on crop production. Crop yields may (Piemental 2012) or may not increase with the application of biochar, depending on soil type and fertilizer management.

The effect of biochar addition on GHGs emissions including CO_2 and N_2O has also been broadly reported with some varied results. Extensive research was focused on the addition of biochar to soils for mitigating N_2O emissions during laboratory or greenhouse incubations under various conditions. The application scopes ranged from soybeans, grass ecosystems (Rondon et al. 2005), common beans (Rondon et al. 2007), rice production (Zhang et al. 2012) or wheat plots (Castaldi et al. 2011) to different agricultural soils (Cayuela et al. 2013). Rondon et al. (2005) found that N_2O emissions were decreased by up to 50% for soybeans and by up to 80% for grasses growing in a low fertility oxisol from the Colombian savanna. Castaldi et al. (2011) cultivated wheat with biochar addition and found that in char treated plots, soil N_2O fluxes were from 26% to 79% lower than N_2O fluxes in the control plots. Similar results were obtained by Zhang et al. (2012), who investigated biochar effects on N_2O emission in rice paddy during a 2-year consecutive field experiment and observed a consistent reduction in N_2O emission in a single crop cycle after biochar amendment.

In contrast to decreases in N_2O emission in most cases, wide variations in the rates on CO_2 emissions from soils treated with biochar have been reported in the literature. For example, Spokas et al. (2009) observed a rate of >20% (w/w) in reduced emission of CO_2 from a silt loam soil amended with wood chip biochar compared to un-amended control. Liu et al. (2011) reported that CO_2 emission was reduced from the waterlogged paddy soil amended with bamboo (*Bambuseae spp.*) and rice straw biochar pyrolyzed at 600 °C. In contrast, Bell and Worrall (2011) observed a significant increase in soil respiration from unplanted plots but not from vegetated plots under lump-wood biochar amendment at the rate of 62.5 t ha^{-1} (approximately 50 t C ha^{-1} input) to an arable soil from Northeast England. Similarly, A 100-day incubation study conducted by Spokas et al. (2009) demonstrated that when three different soil types were amended with 16 types of biochars, three kinds of effects including repression, no change, and stimulation of CO_2 respiration due to biochar addition were observed.

This study was to investigate simultaneous production of biochar and syngas affected by airflow and understand biochar effects on manure composting and associated greenhouse gas emissions.

2 RESEARCH MATERIALS AND METHODS

A lab-scale TLUD gasifier system has been set up. The gasification unit was a 60-inch (152-cm) high and 4-inch (10.1-cm) diameter black iron tube. Air was supplied to the reactor with an air compressor (1.5 kW – 8.62 Bar maximum operational pressure) equipped with a 6-gallon (18.92-liter) reservoir tank (WEN, Elgin, IL). The flow of air was controlled with a variable area flow meter (Cole-Parmer 150-mm, max. pressure 200 psi, Chicago, IL). The temperatures at the top, middle and bottom of the gasifier are recorded with a data logger (Measurement Computing, model: USB-5201, Norton, MA). Tar in the syngas was collected using a two-stage cold trapping method; the first stage contained two flasks cooled by ice in which heavy tar and water are collected. In the second stage, two flasks under dry ice (solid carbon dioxide) cooled the syngas and condense the remaining tar in the gas mixture. Two feedstocks including pine woodchips and rice hulls were studied.

Chicken manure and rice hull biochar were pre-mixed and co-composted to produce an organic fertilizer (Jia et al. 2016). The co-composting experiments were conducted at 35 °C ambient temperature in enclosed reactors where aeration was provided by turning. Various contents of biochar were studied, ranging from 0 to 20% by weight (all treatments also contained 10% sawdust with the rest being manure). The control contained only sawdust and manure but no biochar. The process continued for 6 weeks, and N transformation and greenhouse gas emissions were monitored and are shown below.

3 RESULTS

As in Figure 1, biochar yields of the two biomasses were found to depend on airflow rate, e.g., all decreased as the airflow rate increased (James et al. 2018). Rich hulls had significantly higher biochar yield than woodchips at all airflow rates.

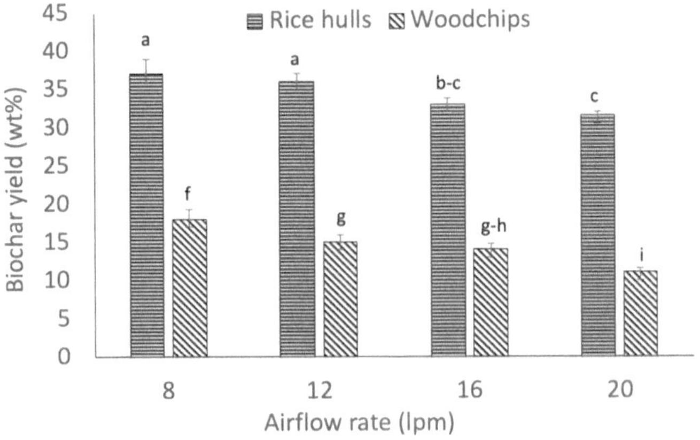

Figure 1. Biochar yield of rice hulls and woodchips at varying airflow rates (James et al. 2018).

Table 1. Syngas composition of rice hulls and woodchips at varying airflow rates (James et al. 2018).

	Airflow (lpm)	H_2 (%)	CO (%)	CO_2 (%)
Rice Hulls	8	2.83±0.30	14.22±0.56	12.86±0.11
	12	3.69±0.46	15.09±0.73	12.50±0.51
	16	4.26±0.17	15.97±0.06	11.83±0.13
	20	4.44±0.13	15.80±0.21	12.22±0.06
Wood chips	8	3.31±0.21	13.72±0.28	13.36±0.19
	12	4.68±0.17	14.27±0.37	13.35±0.87
	16	5.43±0.07	14.23±0.29	13.27±0.41
	20	6.61±0.38	14.97±0.22	13.48±0.68

The contents of H_2 and CO in syngas were affected by airflow rates and insulation conditions (Table 1) (James et al. 2018). Hydrogen production was stimulated by increased airflow; the highest H_2 concentrations were observed at 20 lpm airflow with insulation. Slight increases were noticed in CO content with increasing airflows (James et al. 2018).

Biochar addition had significant effects on NH_4^+-N retention as shown in Figure 2B (Jia et al. 2018). More biochar resulted in higher NH4+-N retention rates in the final product at the end of the co-composting process. Similar results were found for NO_3^-N. Biochar addition also

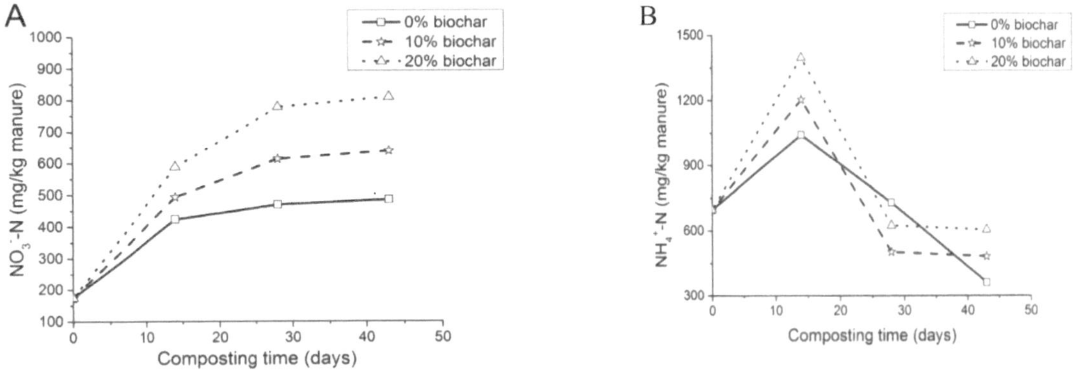

Figure 2. The effect of biochar on (A) NO_3–N and (B) NH^{4+}-N retention in biochar-manure co-composting (Jia et al. 2016).

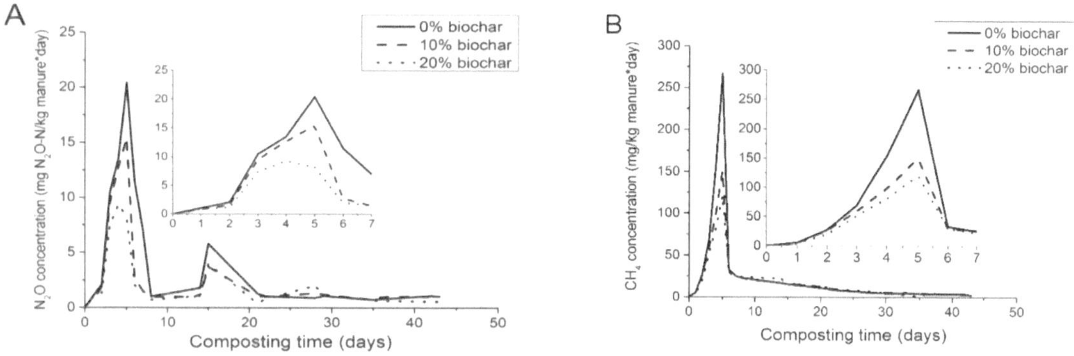

Figure 3. The effect of biochar on (A) N_2O and (B) CH_4 emissions in biochar-manure co-composting (Jia et al. 2016, 2018).

significantly increased NO_3^- -N in the final product (Figure 2A). It is also interesting to note N_2O and CH_4 emission reductions with biochar addition (Figure 3) (Jia et al. 2016).

4 DISCUSSION AND CONCLUSIONS

Simultaneous production of biochar and combustible syngas was achieved with top-lit updraft gasification of rice hulls and woodchips. Biochar yield decreased with increasing airflow rate, while H_2 and CO contents in syngas increased with higher airflow. Adding biochar to chicken manure composting significantly reduced N_2O and CH_4 emissions, which also increased nitrogen retention in the final compost as an organic fertilizer or soil nutrient carrier.

Future research will be focused on understanding the effect of gasification conditions on biochar properties, and correlating biochar properties with soil application related greenhouse gas emissions and soil properties, as well as crop performance.

ACKNOWLEDGEMENT

This study was supported by the USDA National Institute of Food and Agriculture, NIFA Grant 12690802 and Hatch Project NC02613.

REFERENCES

Asai, H., B. Samson, H. Stephan, K. Songyikhangsuthor, K. Homma, Y. Kiyono, Y. Inoue, T. Shiraiwa, and T. Horie. 2009. Biochar amendment techniques for upland rice production in northern laos 1. soil physical properties, leaf SPAD and grain yield. *Field Crops Res.* 111:81–84.

Bell, M., and F. Worrall. 2011. Charcoal addition to soils in NE England: a carbon sink with environmental co-benefits. *Science of the Total Environment 409*(9):1704–1714.

Birzer C., P. Medwell, J. Wilkey, T. West, M. Higgins, and G. MacFarlane. 2013. An analysis of combustion from a top-lit up-draft (TLUD) cookstove. *Journal of Humanitarian Engineering* 2(1):1–7.

Castaldi, S., M. Riondino, S. Baronti, F. Esposito, R. Marzaioli, F. Rutigliano, and F. Miglietta. 2011. Impact of biochar application to a Mediterranean wheat crop on soil microbial activity and greenhouse gas fluxes. *Chemosphere* 85(9):1464–1471.

Cayuela, M., M. Sánchez-Monedero, A. Roig, K. Hanley, A. Enders, and J. Lehmann. 2013. Biochar and denitrification in soils: when, how much and why does biochar reduce N2O emissions? *Scientific reports*, 3:1732–1738.

Chovancova J., Vavrek R., 2019. Decoupling analysis of energy consumption and economic growth of V4 countries. *Problemy Ekorozwoju/Problems of Sustainable Development* 14(1):159–165.
Demirbas, A. 2004. Effects of temperature and particle size on bio-char yield from pyrolysis of agricultural residues. J. Anal. Appl. Pyrolysis 72:243–248.
Dowbor, L. 2013. Economic Democracy – meeting some management challenges: changing scenarios in Brazil. *Problemy Ekorozwoju/Problems of Sustainable Development* 8(2):17–25.
Dunn, B., D. Covington, P. Cole, R. Pugmire, H. Meuzelaar, R. Ernst, E. Heider, and E. Eyring. 2004. Silica xerogel supported cobalt metal Fischer-Tropsch catalysts for syngas to diesel range fuel conversion. *Energy and Fuel* 18(5):1519–1521.
Gaskin, J., R. Speir, K. Harris, K. Das, R. Lee, L. Morris, and D. Fisher. 2010. Effect of peanut hull and pine chip biochar on soil nutrients, corn nutrient status, and yield. *Agron. J.* 102:623–633.
James, A., W. Yuan, M. Boyette, and D. Wang. 2018. Airflow and insulation effects on simultaneous syngas and biochar production in a top-lit updraft biomass gasifier. *Renewable Energy*. 117:116–124.
Jia, X., M. Wang, W. Yuan, X. Ju, and B. Yang. 2016. The influence of biochar addition on chicken manure composting and associated methane and carbon dioxide emissions. *Bioresources* 11(2):5255–5264.
Jia, X., M. Wang, W. Yuan, S. Sha, W. Shi, X. Ju, and B. Yang. 2016. N_2O emission and nitrogen transformation in chicken manure and biochar co-aging. *Transaction of the ASABE* 59(5):1277–1283.
Karhu, K., T. Mattila, I. Bergström, and K. Regina. 2011. Biochar addition to agricultural soil increased CH 4 uptake and water holding capacity–results from a short-term pilot field study. *Agriculture, Ecosystems & Environment* 140(1):309–313.
Knoblauch, C., Marifaat, A.A., Haefele, M.S., 2008. Biochar in rice-based system: impact on carbon mineralization and trace gas emissions. http://www.biocharinternational.org/2008/conference/posters.
Kumar, P., Sharma H., Pal, N., Sadhu, P.K. 2019. Comparative assessment and obstacles in the advancement of renewable energy in India and China. *Problemy Ekorozwoju/Problems of Sustainable Development* 14(2):191–200.
Liu, Y., M. Yang, Y. Wu, H. Wang, Y. Chen, and W. Wu. 2011. Reducing CH4 and CO_2 emissions from waterlogged paddy soil with biochar. *Journal of Soils and Sediments* 11(6):930–939.
Piemental D. 2012. Energy production from Maize. *Problemy Ekorozwoju/Problems of Sustainable Development* 7(2):15–22.
Rondon, M., J. Ramirez, and J. Lehmann. 2005. Charcoal additions reduce net emissions of greenhouse gases to the atmosphere. In: Proceedings of the 3rd USDA Symposium on Greenhouse Gases and Carbon Sequestration, Baltimore, USA, March 21-24, p. 208.
Rondon, M., J. Lehmann, J. Ramírez, and M. Hurtado. 2007. Biological nitrogen fixation by common beans (Phaseolus vulgaris L.) increases with bio-char additions. Biology and Fertility of Soils 43(6):699–708.
Shaw, K. Implementing sustainability in global supply chain. *Problemy Ekorozwoju/Problems of Sustainable Development* 14(2):117–127.
Spokas, K. A., Koskinen, W. C., Baker, J. M., & Reicosky, D. C. (2009). Impacts of woodchip biochar additions on greenhouse gas production and sorption/degradation of two herbicides in a Minnesota soil. *Chemosphere*, 77(4), 574–581.
Tryner J., B. Willson, and A. Marchese. 2014. The effects of fuel type and stove design on emissions and efficiency of natural-draft semi-gasifier biomass cookstoves. *Energy for Sustainable Development* 23:99–109.
Zhang, A., R. Bian, G. Pan, L. Cui, Q. Hussain, L. Li, and X. Yu. 2012. Effects of biochar amendment on soil quality, crop yield and greenhouse gas emission in a Chinese rice paddy: a field study of 2 consecutive rice growing cycles. *Field Crops Research* 127:153–160.

The Role of Agriculture in Climate Change Mitigation – Pawłowski, Litwińczuk & Zhou (eds)
© 2020 Taylor & Francis Group, London, ISBN 978-0-367-43372-7

Forest and climate change – a global view and local cases

M. Pietrzykowski, B. Woś, J. Likus-Cieślik, B. Świątek & M. Pająk
Department of Ecology and Silviculture, Faculty of Forestry, University of Agriculture in Kraków, Kraków, Poland

1 FOREST IN THE WORLD - GLOBAL VIEW

It is estimated that in 2015 forests occupied a total area of about 3,999 million ha, which constituted approximately one third of land on the Earth. Tropical forests (about 1,770 million ha), followed by boreal forests (incl. polar forests) (1,224 million ha) occupied the largest area. The temperate forests covered an area of approximately 685 million ha, and the smallest area (320 million ha) was occupied by the subtropical forests (Keenan et al. 2015). The total forest area reduced compared to 1990 estimates (4,128 million ha); a decrease of approximately 129 million ha was recorded from 1990 to 2015, and an annual decrease of 7.6 million ha occurred in the period from 2010 to 2015. The average forest area per capita decreased from 0.8 ha of forest per person in 1990 to 0.6 ha of forest per person in 2015. The largest decrease in the areas occupied by forests was recorded in the tropical and subtropical climate zones and to a smaller extent in other climate zones. Only in the temperate climate zone, the area occupied by forests has been gradually growing (FAO 2016).

2 CLIMATE CHANGE

The combustion of fossil fuels is the main anthropogenic cause of the progressing climate change and an increase in the concentration of greenhouse gases, mainly carbon dioxide. The carbon dioxide emissions from this source were 36.2 $GtCO_2$ in 2017, and in 2018 increased to 37.1 $GtCO_2$ (Le Quéré et al. 2018). The largest emitters of greenhouse gases from the combustion of fossil fuels are China, the United States, the EU-28, India, Russia and Japan, which together account for 68% of the global CO_2 emissions. In 2017, the EU-28 CO_2 emissions amounted to 3.5 $GtCO_2$; however, they have decreased by 0.9% over the past 5 years, and by 19.5% since 1990 (Muntean et al. 2018).

Deforestation is another anthropogenic cause of a rise in the carbon dioxide concentration. The share of CO_2 emissions due to deforestation and forest degradation in the global greenhouse gas emissions is estimated at around 6-20% depending on the author (van der Werf et al. 2009). As already mentioned, this applies mainly to the tropical and subtropical climate zone. For example, as a result of the tropical forest felling in Brazil and Sumatra, the world's forest cover decreases by about 9.4 million ha per year. Deforestation and transformation of tropical forests into agricultural ecosystems causes the emissions to the atmosphere of about 1.6-1.7 Pg C per year (Sullivan & O'Keeffe 2011, Vaughn 2010).

Progressive climate change may lead to a number of adverse environmental changes. It is estimated that the human activity has increased the Earth's temperature by an average of 1°C (probably from 0.8 to 1.2°C) since the pre-industrial times and by about 0.2°C per decade (Allen et al. 2018). The increase in the average global annual air temperature may be accompanied by unevenly distributed changes in the annual amounts of precipitation and a higher frequency of extreme weather events such as hurricanes (Räisänen et al. 2004).

Some models predict an increased occurrence of droughts in the densely populated areas of Europe, the eastern USA and south-eastern Asia and Brazil in the coming decades, which will have a negative impact on the global economy (Dai 2012).

3 THE ROLE OF FORESTS IN MITIGATING CLIMATE CHANGE

Forests play an important role in reducing the greenhouse gas (GHG) emissions and they store some of the largest carbon stocks compared to other terrestrial ecosystems (Poeplau et al. 2011). It is estimated that total destruction of forests on our planet would cause an increase of about 200 ppm in the CO_2 concentration in the atmosphere by the end of the 21st century. The forest ecosystems around the world contain about 1,240 Pg of carbon, and its resources vary greatly depending on the latitude (Lal 2005). It is estimated that tropical forests accumulate a total of 210 to 238 Pg C (122-136 Mg ha^{-1}) in soil and from 212 to 340 Pg C (121-157 Mg ha^{-1}) in vegetation, while boreal forests from 338 to 471 Pg C (296-378 Mg ha^{-1}) in soil and from 57 to 98 Pg C (53-72 Mg ha^{-1}) in vegetation. Out of these biomes, the lowest carbon stock is accumulated by the temperate zone forests where the total C content in soil ranges from 100 to 153 Pg C (96-122 Mg ha^{-1}), and from 59 to 139 Pg C (57-96 Mg ha^{-1}) in vegetation (Dixon et al. 1994, Lal 2005, Hui et al. 2015). In relation to the total C content in forest ecosystems, boreal forests accumulate about 37% C, temperate forests 14% C, and tropical forests 49% C (Lal 2005).

The carbon stocks are decreasing globally as a result of a reduction in the size of forests; however, the carbon stock per hectare remained practically stable in the period from 1990 to 2010. According to these estimates, the world's forests are therefore a net source of emissions due to a decrease in the total forest area. The data availability and quality have improved since FRA 2005, but the reasons for concern still remain. As with growing stock and biomass, the trend data are weak due to the fact that most countries only have national data on growing stock for one point in time. This means that the changes in stocks merely reflect the changes in forest area. Default carbon values for dead wood were omitted from the 2006 IPCC Guidelines and the default values on carbon in litter are very rough. In the case of soil carbon, there are some issues related to the data from the countries that estimate carbon at different soil depths. Lastly, some countries with large areas of forested peat land have had difficulties assessing the soil carbon using the IPCC guidelines (FAO 2010).

The world's forests store 44% of carbon in the biomass, 11% in dead wood and litter, as well as 45% in the soil (FAO 2010). The amount of carbon accumulated in the soil and biomass varies depending on the climate zone (Prentice 2001). The carbon in boreal zones is stored mainly in soil organic matter, while in tropical forests it is distributed between vegetation and the soil (Prentice 2001). Soil organic matter accumulates in cooler climates because it is produced faster than it can be decomposed; in warmer climates, the decomposition process of soil organic matter and the nutrient cycle are rapid (Prentice 2001). Moreover, the research cited by Hui et al. (2015) have indicated that the forests in boreal climate zones had greater capacity for carbon sequestration in soil than the forests in boreal zones. These results have shown that the climate zone and age of afforestation were the dominant factors impacting carbon sequestration, and jointly accounted for 51–63% of the variation in Net ecosystem production (NEP), gross primary production (GPP) and ecosystem respiration (RE). Various forms of forest management may modify the SOC resources (Jandl et al. 2007). It remains unclear to what extent may the intensity and frequency of biomass harvesting be harmful to SOC (Lindner & Karjalainen 2007, Jandl et al. 2007). Any activity that affects the amount of biomass in vegetation and the soil may cause carbon sequestration or a release of carbon into the atmosphere. According to the research, depending on the intensity of deforestation, forest soils may lose from 5 to 17 Tg C per year, which constitutes approx. 57% of total carbon accumulation (Achat et al. 2015).

Table 1. Carbon stock in forests by region and sub-region in 2010, according to FAO (2010).

Region/sub-region	C in biomass		C in dead wood and litter		C in soil		Total C stock	
	Million tonnes	t/ha	Million tonnes	t/ha	Million tonnes	t/ha	Million tonnes	t/ha
Eastern and Southern Africa	15 762	58.9	3 894	14.6	12 292	46.0	31 955	119.4
Northern Africa	1 747	22.2	694	8.8	2 757	35.0	5 198	66.0
Western and Central Africa	38 349	116.9	3 334	10.2	19 464	59.1	61 089	186.2
Total Africa	55 859	82.2	7 922	11.7	34 461	51.1	98 242	145.7
East Asia	8 754	34.4	1 836	7.2	17 270	67.8	27 860	109.4
South and Southeast Asia	25 204	85.6	1 051	3.6	16 466	55.9	42 722	145.1
Western and Central Asia	1 731	39.8	456	12.6	1 594	36.6	3 871	89.0
Total Asia	35 689	60.2	3 434	5.8	35 330	59.6	74 453	125.7
Europe (excl. Russia)	12 510	63.9	3 648	18.6	18 924	96.6	35 083	179.1
Total Europe	45 010	44.8	20 648	20.5	96 924	96.4	162 583	161.8
Caribbean	516	74.4	103	14.8	416	60.0	1 035	149.2
Central America	1 763	90.4	714	36.6	1 139	58.4	3 616	185.4
North America	37 315	55.0	26 139	38.5	39 643	58.4	103 097	151.8
Total North and Central America	39 594	56.1	26 956	38.2	41 198	58.4	107 747	152.7
Total Oceania	10 480	54.8	2 937	15.3	8 275	43.2	21 692	113.3
Total South America	102 190	118.2	9 990	11.6	75 473	87.3	18 7654	217.1
World	288 821	71.6	71 888	17.8	291 662	72.3	65 2371	161.8

The amount of carbon accumulated in forest ecosystems may be increased through afforestation or reduced through deforestation (Van der Werf et al. 2009; Poeplau et al. 2011, Wei et al. 2014). There are two strategies which take into account the correlation between the carbon dioxide emissions and forest management. The first strategy assumes a long-term accumulation of carbon by forests as old forests are active carbon sinks. In turn, the second strategy assumes an intensification of timber harvesting and using it to replace fossil raw materials in energy production. Increasing the tree biomass can help replace the fossil energy with firewood. It should be emphasised that a characteristic feature of biomass combustion is zero carbon dioxide emission into the atmosphere, which is not the case for fossil fuels (Ellison et al. 2017). It is estimated that about 31% of the world's forests in 2015 were designated as managed forests. The acquisition of wood raw material is gradually increasing and it is reported that in 2011 it amounted to 3.0 billion m^3. From 1990 to 2015, an area of 105 million ha was afforested. From 2000 to 2005, afforestation reached a record level and amounted to about 5.9 million ha of forest per year, while between 2010 and 2015 it fell to 3.3 million ha (FAO 2016).

The anthropogenic activity, including land use change, has a huge impact on the temperature rise and the changes in water availability (Steffen et al. 2015). Forests play a key role in the circulation of atmospheric humidity. The humidity from evapotranspiration may promote and intensify the water redistribution on land (Ellison et al. 2017). On average, about 40% of rainfall on land comes from evapotranspiration, whereas in the Amazonian forests, evapotranspiration is responsible for up to 70% of rainfall (Van der Ent et al. 2010), and a 10% increase in relative humidity can lead to a threefold increase in rainfall (Fan et al. 2007, Khain 2009). An analysis of satellite photos shows that the forests in Europe have a large impact on cloud formation which translates into the momentum of insolation and precipitation (Debortoli et al.

2017, Teuling et al. 2017). Furthermore, through volatile organic compounds, trees contribute to increased rainfall; they also retain and increase the rainwater infiltration (Neary et al. 2009).

Tree felling and the transformation of forest land into arable fields reduce the rainwater infiltration (Nyberg et al. 2012, Zimmermann & Elsenbeer 2008).

Depending on the amount of rainfall, tree roots can collect water in the rainless periods and support water infiltration in the rainy periods (Neumann & Cardon 2012, Prieto et al. 2012). As a result of soil penetration by root systems and of litter accumulation, soil structure is improved which increases its water storage capacity.

Table 2. Carbon stock in living forest biomass in Europe in 2010 and 2015 according to FAO (2016).

Country	Million tonnes in 2010	Million tonnes in 2015	t/ha in 2015
Malta	0	0	173
Czech Republic	323	366	137
Croatia	221	256	133
Switzerland	142	152	121
Slovenia	107	141	113
Slovakia	190	218	112
Luxemburg	9	9	108
Belgium	60	72	106
Germany	1043	1189	104
Austria	368	391	101
Romania	383	616	90
Poland	546	822	87
Serbia	138	237	87
Netherlands	24	32	86
Latvia	231	285	85
Ukraine	662	783	81
France	1049	1364	80
Lithuania	146	167	77
Belarus	482	646	75
Moldova	26	31	75
UK	175	237	75
Estonia	158	165	74
Lichtenstein	0	0	69
Turkey	604	808	69
Italy	496	641	69
Denmark	37	41	67
Albania	49	50	64
Macedonia	62	60	61
Hungary	107	122	59
Bulgaria	161	213	56
Bosnia and Herzegovina	118	118	54
Russia	32157	32800	40
Sweden	1016	1114	40
Norway	377	476	39
Finland	716	780	35
Spain	454	610	33
Cyprus	3	4	22
Greece	73	82	20
Iceland	0	1	13
Ireland	34	n.d	n.d
Portugal	109	n.d.	n.d

n.d. – no data available

4 FOREST RESPONSE TO CLIMATE CHANGE

The adaptation of trees to climate change is very important in the context of the occurrence of extreme weather conditions (Aitken et al. 2008). Periodic or permanent threats to forests caused by a number of abiotic, biotic and anthropogenic factors affect the tree growth both directly and indirectly by reducing photosynthesis and increasing the susceptibility to insect and pathogenic infections (Taeger et al. 2013). On the other hand, increased access to nitrogen and carbon dioxide along with extended vegetation periods and higher temperatures causes an acceleration of the biomass increment while reducing the natural tree mortality. However, very rapidly growing and highly concentrated stands are particularly exposed to the drought stress (Rebetez & Dobbertin, 2004).

The phenotypic and genotypic response of species to a changing climate includes a decrease in the content of nitrogen and phosphorus in the leaves, which is more marked towards the equator where the average annual air temperature and length of the growing season increase. Such a response demonstrates the adaptability and acclimatization of trees to a changing environment (Chapin et al. 1983, Oleksyn et al. 1998, Oleksyn et al. 2002, Woods et al. 2003). As a result of a progressing climate change, it is expected that the range of some tree species will shift. Additionally, the climate change may lead to a horizontal shift in the treeline in the far north and to a vertical one in mountainous regions (Brooker et al. 2007).

According to some forecasts, the forests in northern Europe may show higher productivity due to the climate change, and the species composition may change due to the range shifts. In turn, the southern and western Europe is facing a major challenge due to a predicted decline in forest productivity and possible large-scale disintegration of forest ecosystems (Schelhaas et al. 2015). It is predicted that the climate change may cause a drop in forest productivity as a result of the replacement of highly productive stands of coniferous species (*Picea abies* and *Pinus sylvestris*) by stands of less productive species, e.g. oaks (*Quercus* sp.) up to 2100 (Hanewinkel et al. 2013). The models which forecast the effects of the climate change on the shifting ranges of forest trees in the period from 2061 to 2080 have divided the species into "winners" (mostly late-successional species such as *Abies alba, Fagus sylvatica, Fraxinus excelsior, Quercus robur, Quercus petraea*), which will increase their existing range with global warming; "losers" (mostly pioneer and coniferous species such as *Betula pendula, Larix decidua, Picea abies, Pinus sylvestris*), which will reduce their range; and alien species (*Pseudotsuga menzesii, Quercus rubra, Robinia pseudoacacia*), which will also increase their range, similarly to the winners. At the same time, assuming limited migration, most species will reduce their current range. The range loss will most affect the species whose ranges are most extended to the north (Dyderski et al. 2018).

5 EUROPEAN FORESTS AND UNFAVOURABLE PHENOMENA AFFECTING FORESTS CONNECTED WITH CLIMATE CHANGE

In Europe, forests occupied an area of 1,015 million hectares in 2015. This figure was higher than in 1990 (994 million hectares) (Keenan et al. 2015). Coniferous forests dominate in Europe. According to the European Environment Agency (2006), 14 categories of forests occur in Europe (Figure 1):

1. Boreal forests (dominated by *Picea abies, Pinus sylvestris*),
2. Hemiboreal forests and nemoral coniferous and mixed broadleaved coniferous forests (with *Quercus robur, Fraxinus excelsior, Ulmus glabra, Tilia cordata*),
3. Alpine coniferous forests (with *Picea abies, Pinus sylvestris* and *Larix deciduas, Pinus cembra, P. nigra and P. mugo*),
4. Acidophilous oak and oak birch forests (with *Quercus robur, Q. petraea, Betula pendula*),
5. Mesophytic deciduous forests (mixtures *of Carpinus betulus, Quercus petraea, Quercus robur, Fraxinus, Acer spp. and Tilia cordata*),
6. Beech forests (*Fagus sylvatica* or *Fagus orientalis*),

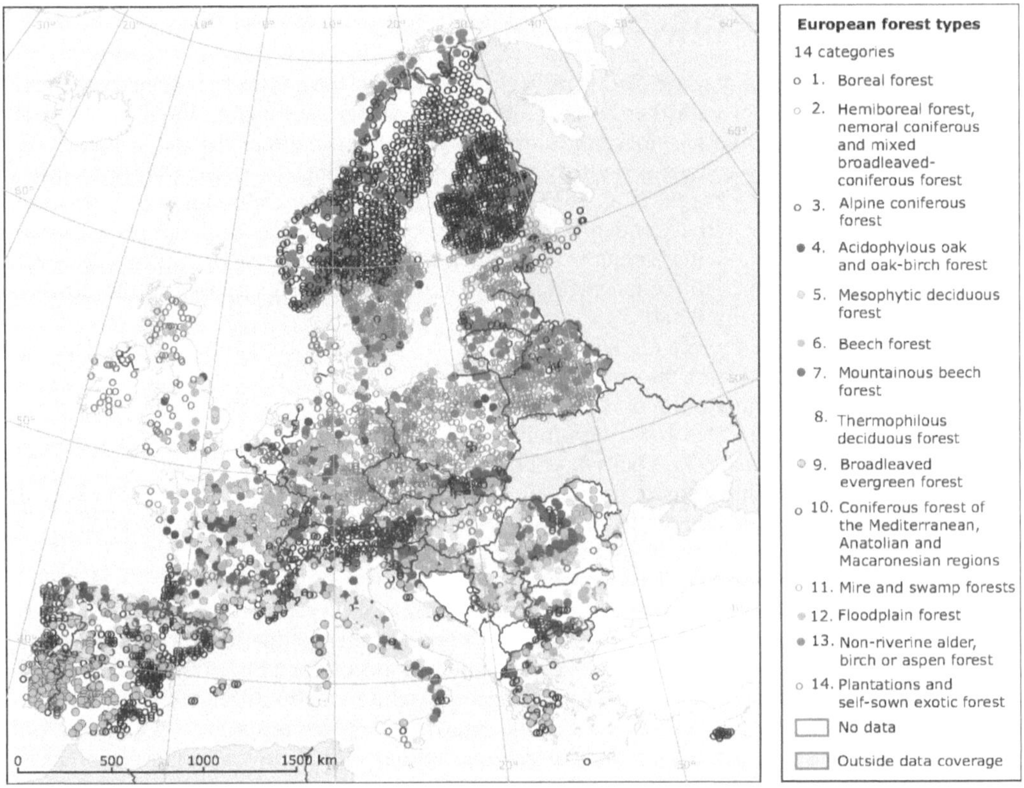

Figure 1. Distribution of individual forest categories in Europe (source: European Environment Agency, 2009).

7. Mountainous beech forests (*Fagus sylvatica, Picea abies, Abies alba, Betula pendula*),
8. Thermophilous deciduous forests (*Quercus* sp., *Acer* sp., *Ostrya* sp., *Fraxinus* sp., *Carpinus* sp.),
9. Broadleaved evergreen forests (broadleaved sclerophyllous or lauriphyllous evergreen trees),
10. Coniferous forests of the Mediterranean, Anatolian and Macaronesian regions (*Pinus* sp., *Abies* sp. and *Juniperus* sp.),
11. Mire and swamp forests (*Picea abies, Pinus sylvestris, Alnus, Betula, Quercus* and *Populus),*
12. Floodplain forests (*Alnus* sp., *Betula* sp., *Populus* sp., *Salix* sp., *Fraxinus* sp., *Ulmus* sp.),
13. Non-riverine alder, birch, or aspen forests (*Alnus* sp., *Betula* sp. or *Populus* sp.),
14. Plantations and self-sown exotic forests.

In the EU, forests and other wooded land constitute over 182 million hectares of land i.e. approx. 43% of the total area. Per capita amounts to 0.36 hectares (Forest Europe 2015). The European forests are an important factor in mitigating the climate change. The European Union (EU) accounts for approximately 5% of the world's forest resources. The impact of forests on the climate was mentioned in the Commission Staff Working Document Multiannual Implementation Plan of the new EU Forest Strategy by the European Commission in 2015 as Priority area n°3: Forests in a changing climate. Forests can help mitigate the climate change, and the associated extreme weather events, and must therefore maintain and enhance their resilience and adaptive capacity (Commission Staff Working Document, 2015). The forests area in Sweden (30.5 million hectares), Spain (27.6 million) and Finland (23.0 million) are among the largest (Forest Europe, 2015).

It is estimated that the total stem volume for European forests is 35 trillion m^3, of which 84% is available for timber production. Over the past 25 years, the total stem volume has grown annually by 403 million m^3. The average growing stock density of European forests is 163 m^3 ha^{-1}. The highest growing stock density has been reported in the Central-Western

Europe region (Germany, France, Switzerland, the Benelux and Great Britain) with 238 m^3 ha^{-1} and Central-Eastern Europe region (Poland, the Czech Republic, Slovakia, Hungary, Romania, Belarus, Ukraine, Moldova) with 247 m^3 ha^{-1} (Forest Europe 2015).

The growing stock of timber in forests and other wooded land in the EU-28 approximated 26.0 billion m^3 (over bark) in 2015. The largest share of this (14.1%) occurred in Germany, followed by Sweden (11.5%) and France (10.0%; Table 2). Germany also had the largest growing stock in forests available for wood supply in 2015, about 3.5 billion m^3. Finland, Poland, Sweden and France each reported between 2.0 and 2.7 billion m^3. The net annual increment (i.e. the average growth in volume of the stock of living trees available at the start of the year minus the average natural mortality of this stock) in the forests available for wood supply was also the highest in Germany, amounting to 119 million m^3 (15.9% of the total increase for the EU-28), while Sweden, France and Finland each accounted for between 10% and 13% of the net annual increment in the EU in 2010, the latest reference year available (Forest Europe 2015).

From 2005 to 2015, the average annual carbon sequestration in the European forest ecosystems amounted to 719 million tonnes CO_2, of which over 50% in forest soils (Forest Europe 2015). It is estimated that in 2017, the European Union forests absorbed an equivalent of 362.8 million tonnes CO_2, which is 8% of total greenhouse gas emissions in the European Union. The largest amounts of greenhouse gases in relation to the total emissions in a given country are absorbed by the forests in Sweden (about 78%), Norway (about 53%), Finland (about 47%), Latvia (about 42.5%) and Lithuania (about 38%) (Forest Europe 2015).

6 REGIONAL SITUATION EXEMPLIFIED BY POLAND – THE IDEA OF CARBON FORESTRY FARMS

The forests in Poland occupy an area of over 9 million hectares (FAO 2017, Eurostat 2018), which represents 30.8% of the total area of the country (FAO 2017). However, according to other estimates made on the basis of spatial data, pursuant to the definition of forests adopted under the Kyoto Protocol, forests cover almost 10.5 million hectares in Poland and thus constitute about 33.5% of the country's area (Hościło et al. 2016). Most forests in Poland are state-owned (81%). Only 19% of forests belong to the private sector (Statistics Poland 2019). The average age of forest stands in Poland is 58 years. The dominant species in the stands is the Scots pine, which accounts for 58.2% of all forests in Poland (Office of Forest Management and Forest Surveying 2019).

The average forest stand growing stock in Poland is higher compared to the average for European forests and it amounts to 283 m^3 ha^{-1} (Office of Forest Management and Forest Surveying 2019). In 2017, 42.7 million m^3 of wood raw material was obtained in Poland. The average harvesting of wood raw material in the last 20 years amounts to about 59% of the current increment. It is higher than in Ukraine (28.5%), Italy (39.2%) and France (47.3%) and much lower than in Sweden (100.8%), Austria (93.5%) and the Czech Republic (85.2%). In the case of Sweden, however, the use that would exceed the increment is not possible, as a large proportion the of forests located in wetlands and mountainous areas have been declared inaccessible for use (nearly 30%) (National Forest Information Centre, 2018).

Compared to these values, in 2017, the Polish forests accumulated about 36.9 million tonnes of CO_2 equivalent, which is about 9% of the total national greenhouse gas emissions (Eurostat 2019). However, more detailed results indicate that the amount of carbon dioxide absorbed by Polish forests may be higher. Other estimates indicate that the amount of CO_2 absorbed by all the forests in Poland is on average 9 t ha^{-1} per year, which corresponds to the absorption of about 81,800 Gg CO_2 per year, i.e. 25% of the anthropogenic greenhouse gas emissions in Poland in 2010 (Gaj 2012). According to a report in Forestry 2017' (GUS 2018), the carbon stock in living wood biomass amounted to 822 million tonnes in 2015.

The data from the National Forest Soil Inventory (NFSI) indicate that the carbon stock in soil horizons (up to 100 cm deep) at locations throughout Poland is 26 t ha^{-1} in the organic horizons and 66 t ha^{-1} in the mineral horizons, on average (Gruba & Socha 2019). These values were lower than in the applicable bibliography on European forests (on average 22.1 t C ha^{-1} in the organic horizon and 108 t C ha^{-1} in the mineral horizon up to 1 meter deep) (De Vos et al. 2015), which the authors explain by a larger share of pine stands in Poland occurring on poor sandy soils (Gruba & Socha 2019). The dominant species in stands have an impact on soil organic carbon stocks. The soils under pine stands contain less carbon than the soils under the coniferous (spruce and fir) and deciduous (oak and beech) species (Gruba & Socha 2019). Locally, in urban and densely populated areas, a significant role may be played by the forests in the reclaimed post-industrial sites. It is estimated that the sequestration potential in such ecosystems is from 1.62 to even 5.64 t C ha^{-1} (Pietrzykowski & Daniels 2014).

In view of these reasons, according to the Polish policy-makers, higher CO_2 absorption from the atmosphere will be possible given proper management. This is to be accomplished by applying the following measures: changes in use, reconstruction of the species composition of stands and developing a multi-strata structure. The Carbon Forestry Programme was launched to investigate these assumptions. The project involves the determination of the role of forests in mitigating the negative effects of climate change in the context of international agreements, increasing the amount of CO_2 absorbed by forest ecosystems, verifying the effectiveness of the existing measures to increase carbon sequestration in forest ecosystems, creating a model for CO_2 absorption by Polish forests, including carbon forestry farms in the CO_2 offset system. On specially designated sites, steps are taken to increase the accumulation of organic carbon. They include modification of the pre-felling to increase the current timber growth in tree stands, change of the felling age, replacement of planting with self-seeding or sowing of seeds – and in the case of planting, performing it in a way that minimises topsoil violation – as well as replacement of clearcutting with complex pruning. It is estimated that owing to these additional measures, Carbon Forestry Farms will absorb almost 1 million tonnes of CO_2 in the course of 30 years, which corresponds to the amount of 37 t ha^{-1} (data from the State Forests Report 2019, https://klimat.lasy.gov.pl). The role of forests in mitigating climate change and carbon sequestration has also been taken into account in the "2030 Environmental Policy".

6.1 *Climate change mitigation by forest management – conclusions*

Most of the climate change models presented by scientists assume further linear warming based on the changes that have taken place in the recent decades (Hansen et al. 2011). However, some astrophysical studies predict that significant cooling may occur in the next few decades (Marcott et al. 2013).The climate change will disturb the forest growth more often as a result of higher incidence of droughts and other abiotic stress factors (Sturrock et al. 2011). In view of such contradictory scientific predictions, it is advisable to base the stand composition on the species with broad ecological spectra, especially as far as the temperature and precipitation are concerned (Hamann & Wang 2006). According to some models which subjectively assume that the current trend in climate change will persist, in the coming 30 to 80 years, the growing conditions of coniferous species – especially the subboreal species which include common spruce as well as European larch, pine and silver birch – will significantly deteriorate; in contrast, the growing conditions of both oak species, beech, sycamore and small-leaved linden will not change (Dyderski et al. 2018, Brooker et al. 2007). Due to the disappearance of Norway spruce, silver fir should be used more widely in afforestation. As pollen analyses and paleobotanical data show, the species with broad ecological spectra have constituted the backbone of Poland's stands over the last millennia, and in that time, historical periods of warming and cooling exhibited higher temperature amplitudes and longer length than they do now (Knapik 2007). Bearing in mind the occurring climate changes, growing of mixed stands should be promoted taking into account various forms of mixed species in order to spread the forest management and silviculture. Due to the climate change and increased

CO_2 absorption by forests, there may be room for forest management in a reduced production cycle (Giorgi et al. 2004). Further studies on the appropriate use of genetic variations may prove to be a way to forest management and counteract changes occurring in the natural environment.

REFERENCES

Achat D.L., Fortin M., Landmann G., Ringeval B., Augusto L. 2015. Forest soil carbon is threatened by intensive biomass harvesting, *Scientific Reports*, 5, Article number: 15991.

Aitken S.N., Yeaman S., Holliday J.A., Wang T., Curtis-McLane S. 2008. Adaptation, migration or extirpation: Climate change outcomes for tree populations. *Evolutionary Applications*, 1: 95–111.

Allen M.R., Dube O.P., Solecki W., Aragón-Durand F., Cramer W., Humphreys S., Kainuma M., Kala J., Mahowald N., Mulugetta Y., Perez R., Wairiu M., Zickfeld K. 2018. Framing and Context. In: Global Warming of 1.5°C. An IPCC Special Report on the impacts of global warming of 1.5°C above pre-industrial levels and related global greenhousegas emission pathways, in the context of strengthening the global response to the threat of climate change, sustainabledevelopment, and efforts to eradicate poverty.

Masson-Delmotte, V., P. Zhai, H.-O. Pörtner, D. Roberts, J. Skea, P.R. Shukla,A. Pirani, W. Moufouma-Okia, C. Péan, R. Pidcock, S. Connors, J.B.R. Matthews, Y. Chen, X. Zhou, M. I. Gomis, E. Lonnoy, T. Maycock, M. Tignor, and T. Waterfield (eds.). In Press.

Brooker R., Travis J., Clark E. J., Dytham C. 2007. Modelling species' range shifts in a changing climate: The impacts of biotic interactions, dispersal distance and the rate of climate change. *Journal of Theoretical Biology*, 245: 59–65.

Chapin F.S., Oechel W.C. 1983. Photosynthesis, Respiration, and Phosphate Absorption by Carex Aquatilis Ecotypes along Latitudinal and Local Environmental Gradients. *Ecology*, 64(4): 743–751.

Commission staff working document. 2015. Multi-annual Implementation Plan of the new EU Forest Strategy 2015. (https://ec.europa.eu/transparency/regdoc/rep/10102/2015/EN/10102-2015-164-EN-F1-1.PDF)

Dai A. 2012. Increasing drought under global warming in observations and models. Nature Climate Change, 3(1): 52–58, doi:10.1038/nclimate1633.

De Vos B., Cools N., Ilvesniemi H., Vesterdal L., Vanguelova E., Carnicelli S. 2015. Benchmark values for forest soil carbon stocks in Europe: Results from a large scale forest soil survey. *Geoderma*, 251–252:33–46.

Debortoli N.S., Dubreuil V., Hirota M., Filho S.R., Lindoso D.P., Nabucet J. 2017. Detecting deforestation impacts in Southern Amazonia rainfall using rain gauges. *International Journal of Climatology*, 37(6): 2889–2900, doi:10.1002/joc.4886.

Dixon R.K., Solomon A.M., Brown S., Houghton R.A., Trexier M.C., Wisniewski J. 1994. Carbon pools and flux of global forest ecosystems. Science, 263(5144): 185–191, doi:10.1126/science.263.5144.185

Dyderski M.K., Paź S., Frelich L.E., Jagodziński A.M. 2018. How much does climate change threaten European forest tree species distributions? *Global Change Biology*, 24(3): 1150–1163. doi.org/10.1111/gcb.13925

Ellison D., Morris C.E., Locatelli B., Sheil D., Cohen J., Murdiyarso D., Gutierrez V., Van Noordwijk M., Creed I.F., Pokorny J., Gaveau D., Spracklen D.V., Tobella A.B., Ilstedt U., Teuling A.J., Gebrehiwot S.G., Sands D.C., Muys B., Verbist B., Springgay E., Sugandi Y., Sullivan C.A. 2017. Trees, forests and water: cool insights for a hot world. *Global Environ. Change*, 43: 51–61, doi.org/10.1016/j.gloenvcha.2017.01.002

European Environment Agency 2009. Classification of the ICP Forests Level 1 plots with respect to main categories of the European Forest Types. European Environment Agency, Copenhagen, Denmark.

Eurostat 2018. Agriculture, forestry and fishery statistics. Statistical Books. Luxembourg: Publications Office of the European Union.

Eurostat 2019. Greenhouse gas emissions by source sector (source: EEA). Last update: 11-06-2019.

Fan J., Zhang R., Li G., Tao W.-K. 2007. Effects of aerosols and relative humidity on cumulus clouds. *Journal of Geophysical Research*, 112(D14204), doi:10.1029/2006JD008136

FAO 2010. Global Forest Resources Assessment 2010. Main report. FAO Forestry Paper 163.

FAO 2016. Global Forest Resources Assessment 2015. How are the world's forests changing? Second edition (http://www.fao.org/3/a-i4793e.pdf).

FAO 2017 Forest Products 2015 (http://www.fao.org/3/a-i7304m.pdf).

Forest Europe 2015. State of Europe's Forests 2015.

Gaj K. 2012. Carbon dioxide sequestration by Polish forest ecosystems. *Forest Research Papers*, 73(1): 17–21.

Giorgi F., Xunqiang B., Pal J. 2004. Mean, interannual variability and trends in a regional climate change experiment over Europe. II: climate change scenarios (2071-2100). *Climate Dynamics*, 23: 839–858.

Gruba P., Socha J. 2019. Exploring the effects of dominant forest tree species, soil texture, altitude, and pHH2O on soil carbon stocks using generalized additive models. Forest Ecology and Management, 447: 105–114. doi:10.1016/j.foreco.2019.05.061

Hamann A., Wang T. 2006. Potential effects of climate change on ecosystem and tree species distribution in British Columbia. *Ecology*, 87: 2773–2786.

Hanewinkel M., Cullmann D.A., Schelhaas M.J., Nabuurs G.J., Zimmermann, N.E. 2013. Climate change may cause severe loss in the economic value of European forest land. *Nature Climate Change*, 3: 203–207.

Hansen J., Sato M., Kharecha P., von Schuckmann K. 2011. Earth's energy imbalance and implications. *Atmospheric Chemistry and Physics*, 11: 13421–13449, doi.org/10.5194/acp-11-13421-2011

Hościło A., Mirończuk A., Lewandowska A. 2016. Determination of the actual forest area in Poland based on the available spatial datasets. *Sylwan*, 160(8): 627–634.

Hui D., Deng Q., Tian H., Luo Y. 2015. Climate Change and Carbon Sequestration in Forest Ecosystems. *Handbook of Climate Change Mitigation and Adaptation*, 1–40. doi.org/10.1007/978-1-4614-6431-0_13-2

Jandl R., Lindner M., Vesterdal L., Bauwens B., Baritz R., Hagedorn F., Johnson D.W., Minkkinen K., Byrne K.A. 2007. How strongly can forest management influence soil carbon sequestration? *Geoderma*, 137: 253–268, doi.org/10.1016/j.geoderma.2006.09.003

Keenan R.J., Reams G.A., Achard F., de Freitas J.V., Grainger A., Lindquist E. 2015. Dynamics of global forest area: Results from the FAO Global Forest Resources Assessment 2015. *Forest Ecology and Management*, 352: 9–20, doi.org/10.1016/j.foreco.2015.06.014

Khain A.P. 2009. Notes on state-of-art investigations of aerosol effects on precipitation: a critical review. *Environmental Research Letters*, 4(15004), doi. org/10.1088/1748-9326/4/1/015004

Knapik R. 2007. Holocene mountain forest communities changes in Poland in the light of pollen research. Studia i Materiały Centrum Edukacji Przyrodniczo-Leśnej 2-3(16) cz.2 (in. polish).

Lal R. 2005. Forest soils and carbon sequestration. *Forest Ecology and Management*, 220: 242–258, doi. org/10.1016/j.foreco.2005.08.015

Le Quéré C., Andrew R.M., Friedlingstein P., Sitch S., Pongratz, J., et al., 2018. Global Carbon Budget 2017, *Earth System Science Data*, 10: 405–448, doi.org/10.5194/essd-10-405-2018

Lindner M., Karjalainen T. 2007. Carbon inventory methods and carbon mitigation potentials of forests in Europe: a short review of recent progress. *European Journal of Forest Research*, 126(2): 149–156, doi: 10.1007/s10342-006-0161-3

Marcott S.A., Shakun J.D., Clark P.U., Mix A.C.A. 2013. Reconstruction of regional and global temperature for the past 11,300 years. *Science*, 339(6124): 1198–1201, doi:10.1126/science.1228026

Muntean M., Guizzardi D., Schaaf E., Crippa M., Solazzo E., Olivier J.G.J., Vignati E. 2018. Fossil CO2 emissions of all world countries - 2018 Report, EUR 29433 EN, Publications Office of the European Union, Luxembourg, ISBN 978-92-79-97240-9, doi:10.2760/30158, JRC113738

National Forest Information Centre Report (Centrum Informacyjne Lasów Państwowych) 2018. Lasy w Polsce, CILP Warszawa, (in polish).

Neary D.G., Ice G.G., Jackson C.R. 2009. Linkages between forest soils and water quality and quantity. *Forest Ecology and Management*, 258: 2269–2281, doi:10.1016/j.foreco.2009.05.027

Neumann R.B., Cardon Z.G. 2012. The magnitude of hydraulic redistribution by plant roots: a review and synthesis of empirical and modeling studies. *New Phytolgist*, 194: 337–352, doi.org/10.1111/j.1469-8137.2012.04088.x

Nyberg G., Bargués Tobella A., Kinyangi J., Ilstedt U. 2012. Soil property changes over a 120-yr chronosequence from forest to agriculture in western Kenya. *Hydrology and Earth System Sciences*, 16: 2085–2094, doi.org/10.5194/hess-16-2085-2012

Office of Forest Management and Forest Surveying Report (Biuro Urządzania Lasu i Geodezji Leśnej) 2019. Wielkoobszarowa inwentaryzacja stanu lasu w Polsce. Wyniki za okres 2014-2018, Sękocin Stary, (in polish).

Oleksyn J., Modrzyński J., Tjoelker M.G., Zytkowiak R., Reich P.B. Karolewski P. 1998. Growth and physiology of Picea abies populations from elevational transects: common garden evidence for altitudinal ecotypes and cold adaptation. *Functional Ecology*, 12: 573–590, doi.org/10.1046/j.1365-2435.1998.00236.x

Oleksyn J., Reich P.B., Zytkowiak R., Karolewski P., Tjoelker M.G. 2002. Needle nutrients in geographically diverse Pinus sylvestris L. populations. *Annals of Forest Science*, 59(1): 1–18, doi:10.1051/forest:2001001

Pietrzykowski M., Daniels W.L. 2014. Estimation of carbon sequestration by pine (Pinus sylvestris L.) ecosystems developed on reforested post-mining sites in Poland on differing mine soil substrates. *Ecological Engineering*, 73: 209–218, doi.org/10.1016/j.ecoleng.2014.09.058

Poeplau C., Don A., Vesterdal L., Leifeld J., Van Wesemael B., Schumacher J., Gesior A. 2011. Temporal dynamics of soil organic carbon after land-use change in the temperate zone–carbon response functions as a model approach. *Global Change Biology*, 17: 2415–2427, doi.org/10.1111/j.1365-2486.2011.02408.x

Prentice I.C. 2001. The Carbon Cycle and Atmospheric Carbon Dioxide. Climate Change 2001: The Scientific Basis IPCC, Cambridge University Press, Cambridge, UK, pp. 183–237.

Prieto I., Armas C., Pugnaire F.I. 2012. Water release through plant roots: new insights into its consequences at the plant and ecosystem level. *New Phytologist*, 193: 830–841, doi:10.1111/j.1469-8137.2011.04039.x

Räisänen J., Hansson U., Ullerstig A., Döscher R., Graham L. P., Jones C., Meier H. E. M., Samuelsson P., Willén U. 2004. European climate in the late twenty−first century: regional simulations with two driving global models and two forcing scenarios. *Climate Dynamics*, 22: 13–31.

Rebetez M., Dobbertin M. 2004. Climate change may already threaten Scots pine stands in the Swiss Alps. *Theoretical and Applied Climatology*, 79 (1-2): 1–9, doi.org/10.1007/s00704-004-0058-3

Schelhaas MJ., Nabuurs GJ., Hengeveld G. Hengeveld G., Reyer Ch., Hanewinkel M., Zimmermann N., Cullmann D. 2015. Alternative forest management strategies to account for climate change-induced productivity and species suitability changes in Europe. *Regional Environmental Change*, 15(8): 1581–1594, doi.org/10.1007/s10113-015-0788-z

State Forests Report 2019, https://klimat.lasy.gov.pl

Statistics Poland 2019. Local Data Bank, (https://bdl.stat.gov.pl/)

Steffen W., Richardson K., Rockström J., Cornell S.E., Fetzer I., Bennett E.M., Biggs R., Carpenter S. R., Vries W., de Wit C.A., de Folke C., Gerten D., Heinke J., Mace G.M., Persson L.M., Ramanathan V., Reyers B., Sörlin S., 2015. Planetary boundaries: guiding human development on a changing planet. Science, 347, doi.org/10.1126/science.1259855

Sturrock R.N., Frankel S. J., Brown A.V., Hennon P.E., Kliejunas J. T., Lewis K.J., Worrall J.J., Woods A.J. 2011. Climate change and forest diseases. *Plant Pathology*, 60: 133–149, doi.org/10.1111/j.1365-3059.2010.02406.x

Sullivan C.A., O'Keeffe J. 2011. Water, biodiversity and ecosystems: reducing our impact. Water Resources Planning and Management. Cambridge University Press, pp.: 117-130, doi.org/10.1017/CBO9780511974304.009

Taeger S., Zang C., Liesebach M., Schneck V., Menzel A. 2013. Impact of climate and drought events on the growth of Scots pine (Pinus sylvestris L.) provenances. *Forest Ecology and Management*, 307: 30–42, doi.org/10.1016/j.foreco.2013.06.053

Teuling A.J., Taylor C.M., Meirink J.F., Melsen L.A., Miralles D.G., van Heerwaarden C.C., Vautard R., Stegehuis A.I., Nabuurs G.J., Vilá-Guerau de Arellano J. 2017. Observational evidence for cloud cover enhancement over western European forests. *Nature Communications*, 8: 14065, doi.org/10.1038/ncomms14065

Van der Ent R.J., Savenije H.H., Schaefli B., Steele-Dunne S.C. 2010. Origin and fate of atmospheric moisture over continents. *Water Resources Research*, 46: 1–12 doi.org/10.1029/2010WR009127

Van der Werf G.R.,Morton D.C., DeFries R.S., Olivier J.G.J., Kasibhatla P.S., Jackson R.B., Collatz G. J., Randerson J.T. 2009. CO2 emissions from forest loss. *Nature Geoscience* 2: 737–738.

Vaughn C.C., 2010. Biodiversity losses and ecosystem function in freshwaters: emerging conclusions and research directions. *BioScience*, 60(1): 25–35, doi.org/10.1525/bio.2010.60.1.7

Wei X., Shao M., Gale W., Li L. 2014. Global pattern of soil carbon losses due to the conversion of forests to agricultural land. *Scientific Reports*, 4:4062, doi: 10.1038/srep04062

Woods H.A., Makino W., Cotner J.B., Hobbie S.E., Harrison J.F., Acharya K., Elser J.J. 2003. Temperature and the chemical composition of poikilothermic organisms. *Functional Ecology*, 17(2): 237–245, doi.org/10.1046/j.1365-2435.2003.00724.x

Zimmermann B., Elsenbeer H. 2008. Spatial and temporal variability of soil saturated hydraulic conductivity in gradients of disturbance. *Journal of Hydrology*, 361(1): 78–95, doi.10.1016/j.jhydrol.2008.07.027

The Role of Agriculture in Climate Change Mitigation – Pawłowski, Litwińczuk & Zhou (eds)
© 2020 Taylor & Francis Group, London, ISBN 978-0-367-43372-7

Cereals role in carbon dioxide absorption in China and Poland

L. Pawłowski, M. Pawłowska, W. Cel, K.Wójcik Oliveira & R. Dzierżak
Faculty of Environmental Engineering, Lublin University of Technology, Lublin, Poland

L. Wang, C. Li & G. Zhou
Zhejiang A and F University, Hangzhou, China

1 INTRODUCTION

Sequestration of atmospheric carbon (C) into plants and soils is a way of partial compensation the anthropogenic emission of carbon dioxide. This emission is still growing, since the beginning of Industrial Revolution, but the significant increase of carbon dioxide (CO_2) emission is observed since the 1960s (Table 1), mainly due to enhancement of combustion of fossil fuels and cement production (FFC&CP), also as the changes in land use (LUC).

Plants and soils are natural sinks for atmospheric CO_2. Plants absorb globally approximately 451 GtCO_2/yr in the photosynthesis process and at the same time emit 435 Gt CO_2/yr in the respiration and dead biomass decomposition process (IPCC 2013). According to Le Quéré et al. (2018) terrestrial ecosystems (TE), including the organisms and their biotopes, are the main global CO_2 sinks. Their responsibility for net CO_2 absorption has been changed over 1960-2016 from 5.1 to 11.0 Gt CO_2/yr (Table 1). The second important natural net CO_2 sink are oceans. The ocean water (OW) absorption has been changed over 1960-2016 from 3.7 to 9.5 Gt CO_2/yr (Table 1). The remaining CO_2 accumulates in the atmosphere causing an increase in its concentration (Table 2).

The increase in the CO_2 concentration in the atmosphere leads to the enhancement both in the absorption of CO_2 in the photosynthesis process (Kimball et al. 2002) and dissolution of CO_2 in ocean water. Hoverer, the increase in the CO_2 content in water leads to its acidification, which will result in a decrease in the CO_2 dissolution (Schmiel et al. 2001).

Reduction of the CO_2 emission from anthropogenic sources is a way to decrease this gas concentration in the atmosphere. Changing the source of energy production, from fossil fuels to alternative fuels, e.g. biomass is considered as a fundamental condition for achieving this goal. But, an excessively one-sided approach of mitigating CO_2 emission from anthropogenic sources, focusing on limiting fossil fuel combustion, may slow down the economic development of many countries (Xu et al. 2018, Chen et al. 2018). Production of energy from renewable sources - especially promoted in the EU - which aims at the mitigation of CO_2 emissions, often leads to the creation of socio-economic problems. Additionally, its effect on CO_2 reduction in many cases is not so big. Negative examples are the production of biodiesel fuel from the oil obtained from coconut palms, grown in Indonesia on the land acquired by burning off tropical forests, and production of ethanol from corn cultivated at the field that were obtained due to conversion of grassland or forests (Fargione et al. 2008, Searchinger et al. 2008). Promotion of the biofuels was based on a simplified analysis and the assumption that the amount of CO_2 emitted during biofuels combustion is equal to the amount absorbed from the atmosphere in photosynthesis process. Although this statement is true, it does not account for additional energy use for cultivation, harvest, and processing the plants into biofuel. Additionally, in order to create a plantation, the natural ecosystems - such as a tropical forests or peatlands - were destroyed. These ecosystems would absorb greater amounts of CO_2 from the atmosphere than croplands (Cao & Cel 2015).

Therefore, an including the primary function of the area that was used for biomass production into the life cycle assessment of biofuels is very important factor influencing the final results.

According to FAO, the area of agricultural land is 37.4% of the total land area (World Bank data, 2017). Thus, the biomass of the agricultural crops grown on this area is a significant sink for the atmospheric CO_2. Cereals occupy the majority of the world's agricultural crops. In

Table 1. Annual mean global anthropogenic CO_2 emissions and net CO_2 sinks (Le Quéré et al. 2018).

	Emissions (Gt CO_2/yr)			Sinks (Gt CO_2/yr)		
Year	FFC&CP	LUC	Sum	TE	OW	Sum
1960-1969	11.4	5.1	16.5	5.1	3.7	8.8
1970-1979	17.2	4.0	21.2	8.8	4.8	13.6
1980-1989	20.2	4.4	24.6	7.3	6.2	13.5
1990-1999	23.1	4.8	27.9	9.2	7.0	16.2
2000-2009	28.6	4.4	33.0	10.6	7.7	18.3
2007-2016	34.5	4.8	39.3	11.0	8.8	19.8
2016	36.3	4.8	41.1	9.9	9.5	19.4
2017	37.0	4.8	41.8	-	-	-
2018	37.1	5.1	42.2	-	-	-

Table 2. Concentration of CO_2 in the atmosphere (Le Quéré et al. 2018).

year	CO_2 concentration ppm	year	CO_2 concentration ppm
1860	286.4	1980	338.7
1870	287.7	1990	354.4
1880	290.8	2000	369.5
1890	294.4	2010	389.9
1900	295.7	2011	391.6
1910	300.1	2012	393.8
1920	303.4	2013	396.5
1930	307.5	2014	398.6
1940	311.3	2015	400.8
1950	311.3	2016	403.3
1960	316.9	2017	407.4
1970	325.7	2018	409.9

Poland alone, in 2018 cereals were the main crops covering an area of 7.8 million ha. In the sown area structure of all the crops, cereals (such as wheat, rye, barley, oats, triticale, maize cultivated for grains, buckwheat, millet, etc.) constituted 72.1%. (Statistics Poland, 2019).

The paper analyzes the annual amount of CO_2 absorbed in above and below-ground parts of the biomass of cereals cultivated in China and Poland, taking into account the specifics of the crops in these countries.

1.1 *The yield of cereals in the global scale*

The cereals are a group of plants that belong to the grasses (*Poaceae* family). There are one of the most frequently cultivated crops in the world due to properties of their grains. World production of cereals grains is growing every year. Compared to 2018, it increased by 1.2% (FAO, 2019). Globally, maize, rice and wheat dominate in terms of acreage and quantity of the produced grains (Table 3).

The cereals are characterized by a high total biomass (shoots and roots) yield per hectare that is an average of 9.44 ± 0.45 Mg ha^{-1} yr^{-1}. It is lower than the yield of grasses cultivated on meadows and pastures (19.80 ± 1.16 Mg ha^{-1} yr^{-1}), but higher than fibrous plants (7.90 ± 1.00 Mg ha^{-1} yr^{-1}), legumes (3.29 ± 0.63 Mg ha^{-1} yr^{-1}), and oil crops (3.05 ± 1.16 Mg ha^{-1} yr^{-1}) (Mathew at al. 2017). Assuming that carbon contains 45% of total biomass (Chirinda et al. 2011), the average amount of carbon absorbed annually in cereals is 4.25 Mg ha^{-1} yr^{-1}.

According to data given by Ghosh et al. (2014) about of 20–30% of total carbon assimilated by cereals is accumulated in the below-ground parts of the plants. Two years experiment

Table 3. World cereal grains production and cropland area in 2017 (FAOSTAT, 2019).

Cereals	Production quantity [mln ton]	Area [mln ha]	Yield [t/ha]
Barley	147.40	47.01	3.14
Buckwheat	3.83	3.94	0.97
Cereals nes*)	7.04	4.42	1.59
Maize (for grains)	1134.75	197.19	5.75
Millet	28.46	31.24	0.91
Oats	25.95	10.20	2.54
Rice	769.66	167.25	4.60
Rye	13.73	4.48	3.06
Sorghum	57.60	40.67	1.42
Triticale	15.56	4.16	3.74
Wheat	771.72	218.54	3.53

*) including inter alia: canagua or coaihua (*Chenopodium pallidicaule*); quihuicha or Inca wheat (*Amaranthus caudatus*); adlay or Job's tears (Coix lacryma-*jobi*); wild rice (*Zizania aquatica*)

conducted by Chirinda et al. (2011) showed that this parameter ranged in wider interval. The share of root biomass in the total cereal biomass ranged from 16-25 % in the case of winter wheat, and 21-35% in the case of spring barley, and the lowest values were observed when the crops were fertilized with inorganic additives.

The below-ground biomass undergoes a slow decomposition in the soil. Carbon released during this process is partially released in form of CO_2 or incorporated into the soil organic matter and microorganisms.

1.2 *Carbon sequestration in cereals in china and poland*

The analysis conducted on the basis the statistical data on the total surface area under the particular cereals cropping (China Statistical Yearbook Database 2018, Statistic Poland 2018a, b) and the mean values of CO_2 assimilated in particular parts of selected cereals: barley, oat, rye and wheat in China, and barley, oat, rye, triticale and wheat in Poland (Table 4 and 5) showed that CO_2 absorption by these crops play the significant role in bio-sequestration of carbon in the country scale (Tables 6 and 7). The data show that the main cereals absorb from the atmosphere 362.6 mln t CO_2 yr^{-1} in China (excluding the rice, that is the main cereals cultivated in this country), and 86.2 mln t CO_2 yr^{-1} in Poland.

In China 201.7 mln t CO_2 yr^{-1} is absorbed in the form of grains, 132.0 mln t CO_2 yr^{-1} in the form of straw and 29.0 mln t CO_2 yr^{-1} in the form of roots of the analyzed cereals (Table 6), while the main cereals in Poland absorb 34.9 mln t CO_2 yr^{-1} in the form of grains, 34.0 mln t CO_2 yr^{-1} in the form of straw and 17.3 mln t CO_2 yr^{-1} in the form of roots (Table 7).

Table 4. The annual amount of CO_2 assimilated in above and below-ground parts of biomass of selected cereals cultivated at 1 hectare of crops in China*).

	Grain	Straw	Roots	Total
	[t CO_2 ha^{-1} yr^{-1}]			
Winter wheat	8.4	5.5	1.1	15.0
Spring wheat	6.6	2.9	0.7	10.3
Rye	4.8	7.0	1.5	1.2
Barley	6.2	4.8	2.2	1.2
Oats	4.4	4.8	1.8	11.0

*) calculated on the basis of data given by Pawłowski et al. (2019)

Table 5. The annual amount of CO_2 assimilated in above and belowground parts of biomass of selected cereals cultivated at 1 hectare of crops in Poland*).

	Grain	Straw	Roots	Total
	[t CO_2 ha^{-1} yr^{-1}]			
Winter wheat	7.3	5.1	1.8	14.3
Spring wheat	5.5	2.9	1.5	9.9
Rye	4.4	7.3	2.9	14.6
Winter barley	6.2	5.1	3.3	14.6
Spring barley	5.5	4.8	3.7	14.0
Winter triticale	5.9	7.2	4.0	17.1
Spring triticale	4.8	6.2	1.8	12.8
Oats	4.4	4.8	2.6	11.8

*) calculated on the basis of data given by Pawłowski et al. (2019)

Table 6. Annual CO_2 sequestration by cereals cultivated in China.

	Grain	Straw	Roots	Total
	[mln t CO_2/yr]			
Winter wheat	188.5	124.3	26.4	339.2
Spring wheat	9.5	4.4	1.1	15.0
Rye	0.7	1.1	0.4	2.2
Winter barley	2.2	1.5	0.7	4.4
Spring oat	0.7	0.7	0.4	1.8
Total	201.7	132.0	29.0	362.6

Table 7. Annual CO_2 sequestration by cereals cultivated in Poland.

	Grain	Straw	Roots	Total
	[mln t CO_2/yr]			
Winter wheat	14.3	9.9	3.7	27.9
Spring wheat	2.6	1.5	0.7	4.8
Rye	3.0	6.2	2.6	11.8
Winter barley	1.1	1.1	0.7	2.9
Spring barley	4.0	3.7	2.9	10.6
Winter triticale	6.6	8.3	4.8	19.7
Spring triticale	1.1	1.1	0.4	2.6
Oats	2.2	2.2	1.5	5.9
Total	34.9	34.0	17.3	86.2

The carbon absorbed in root biomass is deposited in soil. This does not mean that this amount of absorbed carbon undergoes permanent sequestration. But large part of them is transformed into the humic substances.

Grain mainly serves as substrate for food production or as a fodder for animals. Part of it is released to the atmosphere in the course of breathing of people and animals consuming them, and some part is transformed in the organic compounds during the metabolic processes.

In the past, straw was commonly used as a litter in animal husbandry. Then, it was deposited into soil with manure, where it was transformed into humic substances along with roots increasing productivity of the soil (Pan et al. 2009). At present, large amounts of straw are used for energy purposes and very often it is directly combusted in fields.

The microbiological processes occurring in soil lead to the decomposition of organic substances contained in soil, resulting in the partial release of carbon in form of CO_2 into the atmosphere. The rate of organic substances decomposition is dependent mainly on the availability of oxygen, which infiltrates into the soil during the tillage. Therefore, in order to increase the storage of carbon in soil, zero tillage is recommended.

2 CONCLUSIONS

The main cereals, such as wheat, rye, barley and oats (excluding rice) absorb annually 362.6 mln t CO_2 yr^{-1} in China, and 86.2 mln t CO_2 yr^{-1} in Poland. Most of the carbon is stored in aboveground parts of the biomass of cereals, but 29 mln t CO_2 yr^{-1} and 17.3 mln t CO_2 yr^{-1} is absorbed in the roots of the plants cultivated in China and Poland, respectively. This carbon can be incorporated into the soil organic matter and micro-organisms. But it is difficult to determine how much carbon is deposited permanently in soil. The process of carbon sequestration in soil is influenced by many factors related with soil properties, especially the grain size distribution, water and oxygen accessibility, and pH. It also depends on climate conditions, and chemical properties of organic matter, which is transformed in soil.

REFERENCES

Cao, Y. & Cel, W. 2015. Sustainable mitigation of methane emission by natural processes. *Problemy Ekorozwoju/Problems of Sustainable Development* 10(1): 217–121.

China Statistical Yearbook Database, National Bureau of Statistic of China, 2018.

Chirinda, N.1, Olesen, J.E.1 & Porter, J.R. 2011. High Root Biomass FOR cereal crops increases carbon sequestration in organic arable systems. Proceedings of 17th IFOAM Organic World Congress.

FAO 2019. Crop prospects and food situation, Quarterly Global Report, July 2019 http://www.fao.org/3/ca5327en/ca5327en.pdf.

FAOSTAT data 2019, http://www.fao.org/faostat/en/#data/QC, available online Dec. 29.2019.

Fargione, J., Hill, J., Tilman, D., Polasky S., Hawthorne, P. 2008, Land clearing and the biofuel carbon debt. *Science* 319(5867): 1235–1238.

Ghosh P.K., Narendra Kumar, Venkatesh M.S., Hazra K.K., Nadarajan N. Resource Conservation Technology in Pulses. Scientific Publisher (India), New Delhi, 2014.

IPCC 2013. The Fifth Assessment Report of the Intergovernmental Panel on Climate Change 2013: The Physical Science Basis.

Kimball B. A., Kobayashi K., Bindi M. 2002. Responses of agricultural crops to free air CO2 enrichment. *Advances in Agronomy*, 77, 293–368.

Le Quéré, C. et al. 2018, Global Carbon Budget 2017. Earth System Science Data 10: 405–448.

Mathew, I., Shimelis, H., Mutema, M. Chaplot, V. 2017. What crop type for atmospheric carbon sequestration: results from a global data analysis. *Agriculture, Ecosystems & Environment*, 243: 34–46.

Pan G. Smith P., Pan W. 2009, The role of soil organic matter in maintaining the productivity and yield stability of cereals in China, *Agriculture, Ecosystems & Environment* 129(1): 344–348.

Pawłowski L., Pawłowska M., Cel W., Wang L., Li Ch., Mei T., Characteristic of carbon dioxide absorption by cereals in Poland and China, *Gospodarka Surowcami Mineralnymi – Mineral Resources Management*, 35 (1);165–176.

Schmiel, D.S. et al. 2001, Recent patterns and mechanisms of carbon exchange by terrestrial ecosystems. *Nature* 414(8): 169–172.

Searchinger, T. et al. 2008, Use of U.S. croplands for biofuels increases greenhouse gases through emissions from land-use change. *Science* 319: 1238–1240.

Statistic Poland 2018a. Crop production in 2017. Central Statistical Office. Warsaw 2018.

Statistic Poland 2018b. Land use and sown area in 2017. Central Statistical Office. Warsaw 2018.

World Bank data, 2017, https://data.worldbank.org/indicator/AG.LND.AGRI.ZS.

Xu, J. & Li, J. 2018. The tradeoff between growth and environment: evidence from China and the United States. *Problemy Ekorozwoju/Problems of Sustainable Development* 13(1):15–20.

The Role of Agriculture in Climate Change Mitigation – Pawłowski, Litwińczuk & Zhou (eds)
© 2020 Taylor & Francis Group, London, ISBN 978-0-367-43372-7

Mitigation of greenhouse gas emission through anaerobic digestion of livestock waste

Jung-Jeng Su
Department of Animal Science and Technology, Division Chief of Bioenergy Research Center, National Taiwan University, Taipei, Taiwan (R.O.C.)

1 INTRODUCTION

Pork is the largest economic meat source for the Taiwan population, with a per capita pork consumption of about 36.5 kg/year. As such, pig farmers produced biogas production due to the large presence of pig farms in Taiwan and availability of fresh manure for biogas collection. As a result, the collected biogas from pig farms is used for power generation or direct combustion in Taiwan (Su & Chen 2015).

Greenhouse gas (GHG) emissions from the agricultural sector in Taiwan was 2,712 kilotons of CO_2-eq in 2016, accounting only for 0.93% of the country's total GHG emissions (293,125 kilotons of CO_2-eq). From 1990 to 2016, GHG emissions from the agricultural sector show the agriculture sector decreased by 30.66% with an average annual growth rate of −1.40% (TEPA 2019). In 2016, N_2O emission from agricultural soil, livestock manure management, and agricultural waste burning accounted for 28.0, 1.62, and 0.02% of total agricultural GHG emission (4,701 kilotons of CO_2-eq), respectively (TEPA 2019).

The livestock greenhouse gas emission was mainly from anaerobic digestion of manure and wastewater. Traditional anaerobic digestion (AD) is liquid anaerobic digestion using high water content feedstocks such as wastewater and slurry. However, anaerobic digestion can be divided into different types depending on the total solid content (TS), temperature, and ways of feeding the digester (Kothari et al. 2014). Based on the TS content, AD can be classified into liquid anaerobic digestion, which contains less than 15% TS and is applied for the treatment of wastewater (de Laclos et al. 1997), and solid-state anaerobic digestion (SSAD), which contains more than 15% TS (Ge et al. 2016). Liquid AD is a technology that had been used for a long time, while SSAD for the treatment of municipal solid waste was initially installed in Europe and has gradually increased since the 1990s (Baere & Mattheeuws 2010). Compared to SSAD, liquid AD generates a large amount of wastewater as well as sludge production (Pezzolla et al. 2017). In contrast, SSAD generates a lower amount of wastewater and requires less energy for mixing as well as heating (Kothari et al. 2014, Xu et al. 2014).

2 INVESTIGATION OF BIOGAS PRODUCTION FROM ANAEROBIC DIGESTION OF PIGGERY WASTEWATER

2.1 *Waste-water treatment systems in Taiwan in selected pig farms*

The most widespread piggery waste-water treatment system in Taiwan is the three-step piggery waste-water treatment system (TPWT), which includes the stages of solid/liquid separation, anaerobic digestion and activated sludge treatment (Figure 1). The anaerobic digestion basin is a plug-flow, top-opened, horizontal and underground waste-water basin covered with a plastic lid and constant pressure device: biogas can be collected from the top of the plastic cover. Among this literature, the TPWT system is the only typical wastewater treatment system for manure management applied in Asia (Su et al. 1997).

Three integrated pig farms were selected from Miaoli (9,000-pig farm), Changhua (15,000–-1,8000-pig farm) and Tainan (10,000-pig farm) Counties located in northern, central and

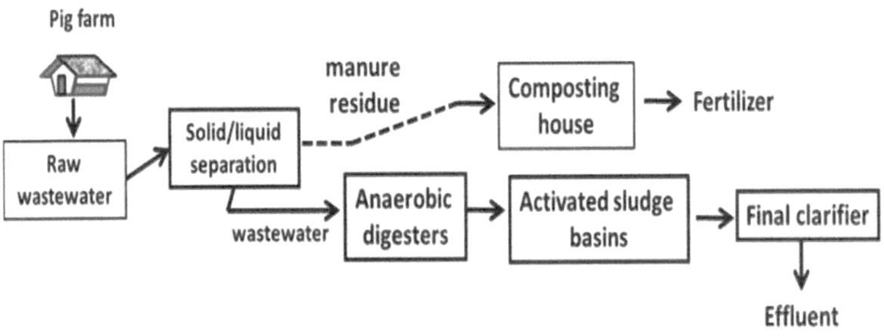

Figure 1. Diagram of the three-step piggery wastewater treatment (TPWT) system (Su and Chen, 2018).

southern Taiwan. The daily wastewater volume was 294, 300, and 400 m^3/d, respectively. Thus, the hydraulic retention time (HRT) of anaerobic digesters was 6.5, 20.4, and 3.0 d, respectively. Because the Tropic of Cancer cuts across central Taiwan, dividing Taiwan between the tropical and subtropical zone, with unique landscapes and rich natural resources of different climates. Hence, the three pig farms were chosen in northern, central and southern Taiwan in order to represent different climatic regions.

2.2 *Production of biogas after anaerobic digestion of piggery waste water*

The emission factors of GHG were calculated from the data of analyzed on-site samples, taken from the gas outlets of the selected anaerobic piggery wastewater treatment facilities prior to pressure stabilizers in northern, central and southern pig farms (Table 1). Daily average biogas production per farm across the temperature zones ranged from 625 to 958 ($P < 0.05$), 1,851 to 2,129 ($P > 0.05$) and 628 to 696 m^3/day ($P > 0.05$) in the northern, central and southern pig farms, respectively. Results implied that the biogas production rates might reach their maximum when hydraulic retention time (HRT) of anaerobic digesters is >20 days. In contrast, there might be inadequate retention time for biogas production when HRT of anaerobic digesters was only 3 days.

The analytical results demonstrated the average GHG compositions in the biogas for CH_4, CO_2 and N_2O were 0.65 ± 0.035, 0.30 ± 0.011 and 0.0004 ± 0.00021, respectively. Additionally, the average emission levels of CH_4, CO_2 and N_2O were 10.8–19.0, 12.3–25.3 and 0.03–0.09 kg/head/year, respectively (Table 1). Statistical results implied that both average biogas production and GHG contents in biogas were significantly different among three pig farms, except for N_2O. These results might be due to different climates and manure management techniques (slatted vs. unsalted) among the three pig farms in the current study.

Table 1. Average biogas production in the northern, central, and southern pig farms (Su and Chen, 2018).

	Farm locations				
Biogas production	Northern	Central	Southern	Average	P-value
Daily average per farm (m^3/d)	865±162	1926±168	664±85	1,151±653	< 0.001
Average per head (m^3/head/d)	0.088±0.016	0.128±0.011	0.066±0.008	0.094±0.031	< 0.05
CH_4 (kg/head/yr)	13.29±0.27	19.02±0.09	10.82±0.05	14.38±4.21	< 0.001
CO_2 (kg/head/yr)	17.73±0.34	25.27±0.16	12.34±0.16	18.45±6.49	< 0.001
N_2O (kg/head/yr)	0.029±0.016	0.050±0.011	0.086±0.002	0.055±0.029	NS

Data presented as mean ±S.D. NS, not significant.

The results suggested that the average emission factor of CH_4 from anaerobic waste-water treatment of pig farms (14.4 kg/head/year) is lower than that (1–23 kg/head/year in temperate and warm regions) estimated by IPCC (2006) (Su & Chen 2018). This difference may result from differences in the organic concentrations in waste water and liquid manure (or slurry).

However, emission of CH_4 for pig operation in the three climate regions, cool (<15 °C), temperate (15–25 °C) and warm (>25 °C), estimated by IPCC in 1996 was 1, 4 and 7 kg/head/year, respectively (IPCC 1996).

3 INVESTIGATION OF BIOGAS PRODUCTION FROM SOLID-STATE ANAEROBIC DIGESTION OF CATTLE MANURE

3.1 *Inoculum for solid-state anaerobic digestion (SSAD) and SSAD reactor design*

Dairy cattle manure from the solid/liquid separation of wastewater from the National Taiwan University (NTU) dairy farm was used to as the sole substrate for SSAD reactors. The sludge from the anaerobic digesters of the NTU dairy farm was used as the initial inoculum for enrichment of the SSAD reactors. After three batches of the enrichment process (about 90 days), the methanogenically activated (MeA) mixture from the previous batches of SSAD were utilized as the inocula for the further SSAD experiments (Figure 2). The initial and final weight of dairy cattle manure and MeA mixture was 5 and 4.1−4.8 kg and the volume of initial and final liquid was 2.5 and 2.7−4.0 L, respectively for SSAD experiment. Acrylic anaerobic digester (19 cm inside diameter× 115 cm height) in triplicate with a working volume of 37 L was used in this study (Figure 3). A thermostatic recirculation water batch was equipped outside the digester to maintain the digester at 36±1°C. Each digester had an independent leachate recirculation magnet pump (Wee & Su 2019), which was used to recirculate the leachate from the side port at the bottom of the digester to the side port at the top of the digester through rubber tubes (18 mm outside diameter × 8 mm inside diameter). A digital programming timer controlled the leachate recirculation and the recirculation frequency was every 20 min for the 20 sec recirculation process (approximately 3.8 L leachate/recirculation).

3.2 *Preliminary study of SSAD by fresh cattle manure*

A preliminary study was performed in an acrylic anaerobic digester (19 cm i.d. × 115 cm Height) with fresh cattle manure (5 kg) as the sole feedstock. The preliminary results showed that two peaks of biogas yield occurred during Days 1 to 7 and Days 30 to 34 (Figure 4A). Nitrogen content in the biogas decreased from 63.8 to 19.2%. However, Methane content increased from 8.5 to 52.7% for the 37-d period. The results implied that denitrification was dominant in the anaerobic digester during Days 1 to 7 and methanogenesis became dominant after Day 16 (Figure 4B) (Wee & Su 2019).

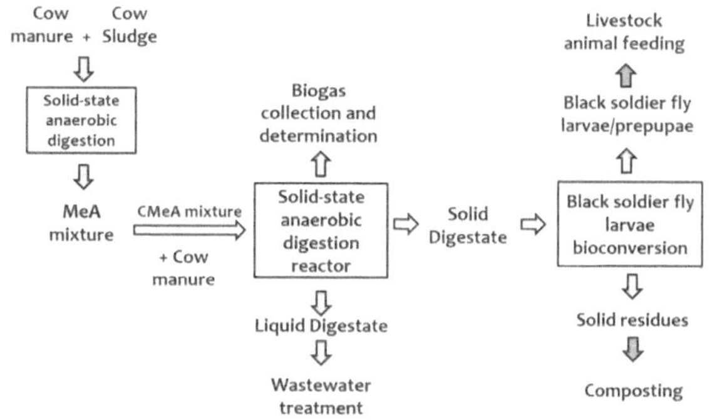

Figure 2. Flow chart of the two-step biological treatment system (Wee & Su 2019).

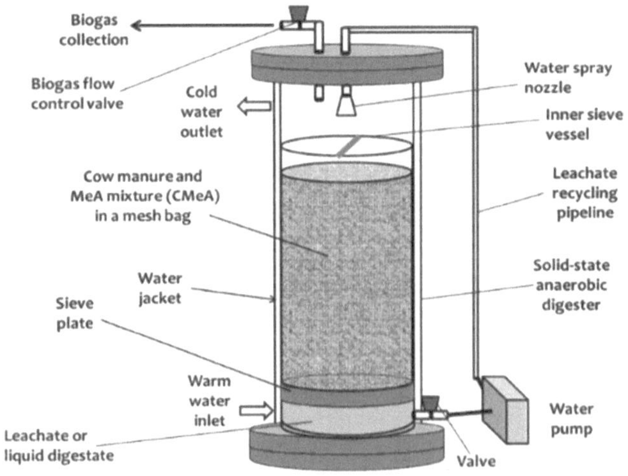

Figure 3. Design of the SSAD reactor (Wee & Su 2019).

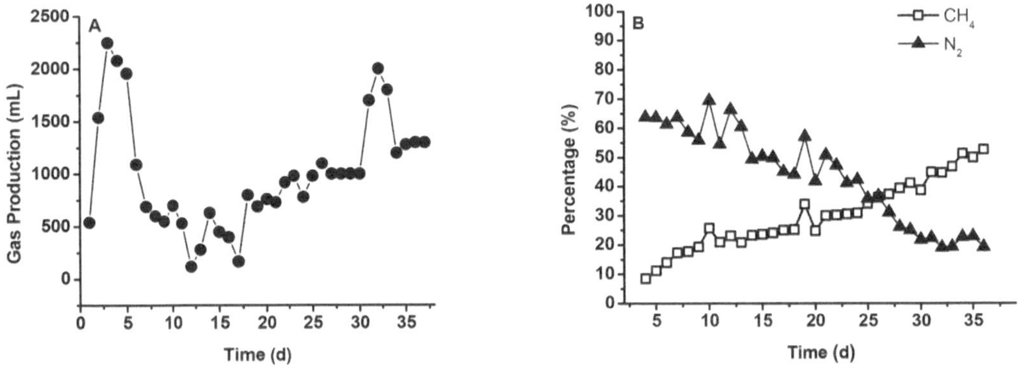

Figure 4. Biogas production (A) and biogas contents (B) of the preliminary SSAD study (Wee & Su 2019).

3.3 *Time course experiment of SSAD*

In order to accelerate the SSAD process, various inoculation ratios (digestate fiber from previous experiments) were tested as well as the operation conditions. Theoretically, the more digestate fiber inoculate, the faster SSAD process (Figure 2). Two groups of initial pH (7.8 and 5.2–5.5) and three groups of inoculation ratios (IR) (50, 30, and 10%) were designed to evaluate efficiency of SSAD experiments. The group with 50, 30, and 10% inoculation ratio are referred to IR_{50}, IR_{30}, and IR_{10}, respectively (Wee & Su 2019). Every time course experiment of SSAD was conducted for a 14-d period, daily biogas yield was collected and measured by applying water displacement method with a 6-L glass gas collector. The MeA mixture from the initial SSAD reactor as inocula were added manually into a laundry mesh bag (60 cm × 60 cm) by the inoculation ratios of 50, 30, and 10% (w/w) respectively, the total weight of the mixture was 5 kg.

3.4 *Effect of pH adjustment and IR on daily and cumulative biogas yield*

Biogas productivity was measured in terms of volatile solids destroyed (VS_{des}) or VS loaded (VS_{load}). Results showed that the group IR_{50} with pH adjustment had the largest biogas peak value, 21.7±0.2 L/kg VS_{load}/day, after the peak of biogas yield on Day 3, the daily biogas yield decreased gradually to 5.81±0.2 L/kg VS_{load}/day on Day 14. While the peak value of daily biogas yield was 17.3±0.3 and 11.7±0.3 L/kg VS_{load}/day for group IR_{30} and IR_{10} with pH adjustment, respectively. The peak value of daily biogas yield was 6.4±0.3, 6.9±0.2, and 4.5 ±0.4 L/kg VS_{load}/day for group IR_{50}, IR_{30}, and IR_{10} without pH adjustment, respectively (Wee and Su, 2019). Experimental results implied that pH was a key parameter that influenced the peak sizes; indeed, the groups with pH adjustment had a larger peak size from 21.7 to 11.7 L/kg VS_{load}/day, where groups without pH adjustment had a small peak size from 6.9 to 4.5 L/kg VSload/day. The cumulative biogas yield of groups with pH adjustment (124.6±5.8 −164.6±1.9 L/kg VS_{load}) were all significantly higher than groups without pH adjustment (37.7±1.9−45.5±2.6 L/kg VS_{load}) (Wee & Su 2019).

3.5 *Effect of pH adjustment and IR on methane concentration*

Methane productivity was measured in terms of VS_{des}, VS_{load}, or volume (Møller et al. 2004). Thus, theoretical methane yield (Bu) and ultimate methane yield (Bo) were defined in terms of either VS destroyed (L CH_4/kg VS_{des}) or VS loaded (L CH_4/kg VS_{load}) based on either the actually bio-degraded or total load VS contents of the substrate mixture, respectively, by the SSAD process. This result indicated the group IR50 with pH adjustment had a faster methanogenesis start-up and resulted in faster increase of methane concentration. The group IR30 with pH adjustment had the highest cumulative methane yield (96.8±2.0 L CH4/kg VS_{load}) on Day 14 than the groups IR10 (69.2±3.7 L CH_4/kg VSload) and IR50 (86.8±1.0 L CH_4/kg VS_{load}) with pH adjustment. However, the groups without pH adjustment had lower cumulative methane yield (12.1±2.5−16.0±1.7 L CH_4/kg VS_{load}) on Day 14 compared to groups with pH adjustment regardless inoculation ratios (Wee & Su 2019).

The study of Møller et al. (2004) showed that the theoretical methane yield and ultimate methane yield of dairy cattle manure were 468±61 L CH_4/kg VS_{des} and 148±41 L CH_4/kg VS_{load}, respectively. The theoretical methane yield of the group IR30 with pH adjustment (626.1±28.7 L CH_4/kg VS_{des}) was comparable to Møller's study (468±61 L CH_4/kg VS_{des}) (Wee & Su 2019, Møller et al. 2004). While the ultimate methane yield of the group IR30 with pH adjustment (96.81±2.0 L CH_4/kg VS_{load}) was slightly lower than Møller's study (148±41 L CH_4/kg VS_{load}). The experimental results implied that the methane productivity of SSAD in this study was comparable to other SSAD studies (Wee & Su 2019).

4 STRATEGY FOR MITIGATION LIVESTOCK GREENHOUSE GAS EMISSION

Some early studies of livestock biogas upgrading and various applications, e.g. power generation, heat lamp, hot water stove, gas burner, absorption chiller, and biogas powered car, was used from 1974 in Taiwan by Taiwan Livestock Research Institute (TLRI) (Hong et al. 1980). However, the study and extension service of biogas utilization were slowed down after 1995 because of insufficient biogas desulfurization resulting in severe corrosion of all related facilities.

Until the year of 2009, a novel biogas bio-desulfurization technique and facility was developed (Su 2017). The latest feasible biogas in situ applications in subtropical and tropical regions such as Taiwan is summarized in Table 2.

4.1 *Biogas power generation*

Field investigation data of the Taiwan Livestock Research Institute (TLRI) showed that the desulfurized biogas produced from the anaerobic digesters of a 20,000-head pig farm was able to operate three sets of 65 kW micro-turbine power generators for producing 2,726 kWh/day in an average of 18.3 h operation daily (Hsiao et al. 2017). Another field investigation data showed that the desulfurized biogas produced from the vertical anaerobic digesters (total

Table 2. Possible biogas in situ applications in subtropical and tropical regions such as Taiwan.

Biogas utilization	Facilities	Output	Relative energy efficiency	Practical/in situ applications
Biogas power generation	Desulfurizer/power generator	Electricity and waste heat	Low/~20% power generation only w/o waste heat recycling	Livestock farm power generation
Biogas heating	Desulfurizer/heating lamp for piglets/waste heat recycling	Heating air or water	High/less chemical energy loss	Animal house heating
Biogas cooling	Desulfurizer/absorption chiller/waste heat recycling	Cooling air or water	High/less chemical energy loss	Animal house cooling
Biogas upgrading	Desulfurizer/CO_2 removal facility	Bio-natural gas	High/less chemical energy loss	Automobile fuel/substitute fuel for thermal power generation
Biogas direct combustion	Desulfurizer/boiler/hot water stove/cooking stoves	Desulfurized biogas	High/less chemical energy loss	Hot water supply and fuel for cooking and shower

volume = 1,200 m^3) of a 10,000-head pig farm was able to operate a 130-kW generator for producing about 1,200 kWh/day in an average of 16 h operation daily (Hsiao et al. 2017). The other field investigation data showed that the desulfurized biogas produced from the vertical anaerobic digesters of a 3,000-head pig farm was able to operate a 50-kW generator for producing about 1,511 kWh/day in an average of 10 h operation daily (Hsiao et al. 2017). Another farm-scale BBS was established and put in operation in 2013 at a 25000-head pig farm, the Central Pig Farm, Pingtung County, Taiwan. The 25000-head pig farm has been selected as a demonstration site for biogas power generation in 2015 by Bureau of Energy, Ministry of Economic Affairs (MEA), Taiwan. About 2500 m^3 of biogas containing 60% methane is produced daily.

Three sets of 65 kW micro-turbine power generators are installed and operated sequentially. Thus, an average of 3,700 kWh electricity is produced daily (Su 2017).

The methane content in the biogas from pig farms is about 60 to 65%. There are about 81 domestic pig farms (total pig number = 490,000) carrying out biogas power generation from 2016 to 2018 in Taiwan based on statistical data of the Council of Agriculture (COA).

4.2 *Biogas heating*

The infrared heat lamp using biogas as the fuel is commonly applied to piglet heating of the pig house in Taiwan. The infrared heat lamp is normally using liquefied petroleum gas (LPG) as the fuel for heating and the optimal LPG/air ratio of 1:23. However, the pore size of the biogas injectors must be enlarged and the volume of inlet air must be reduced to the biogas/air ratio of 1:7.5 for better conditions of biogas combustion. All biogas has to be compressed up to 20 cm/H_2O by using a 1/4 or 1/2 HP biogas compressor prior to heat lamp application (Chow et al. 1981). Up to date, the infrared heat lamp is still a common heating device of Taiwanese commercial pig farms in winter with either biogas or LPG as the fuel.

4.3 *Biogas cooling*

Livestock biogas is seldom used properly. Domestic livestock farms are randomly located that results in difficulty of collecting biogas for any applications. Electricity efficiency of power generation by using biogas is about 20–39%. However, efficiency of absorption chiller for making cool water by directly biogas combustion can be achieved above 80%. The milking records of the dairy farm of NTU showed that there was 16% less average daily milk production from dairy farm during May and October comparing with the average daily milk production vs. thermal humidity index (THI) in other months (Figure 5). Production cost can be reduced, if biogas combustion can be integrated with animal house and personnel cooling, animal heat stress mitigation as well as production efficiency promotion in summer. Biogas can be used for making cold mist to significantly reduce electricity cost besides of power generation in summer. Experimental results of this study applying biogas chiller to produce cold mist for dairy farm cooling in a dairy farm showed that heat stress level on dairy cattle was improved from moderate level to mild level. Application of biogas chiller for keeping dairy cattle house cool is another option for utilizing livestock biogas in summer.

A horizontal absorption freezer (working volume = 150 L, power consumption = 75 W/h, XD-200, Shandong Coner Gas Refrigerator Manufactory Co., Ltd., Shandong, China) was installed next to the anaerobic digesters of the dairy farm of NTU (Figure 6). Desulfurized biogas was used as the single fuel for combustion and heating the core of the freezer. Biogas flow was controlled by gas flow meter (Rate-Master flowmeters 1-5 LPM, Dwyer Instrument, Inc., Michigan City, USA). Temperature was on-line recorded by using mini temperature data logger with sensor (RC-4, range = -40 to +85°C, accuracy = +1°C, IP67, Elitech, China). The methane, nitrogen gas, carbon dioxide, and the others in the biogas of dairy farm of NTU was 61.3±5.2% (55.9 to 62.2%), 14.2 ±6.2% (9.6 to 21.2%), 18.8±1.8% (16.8 to 20.2%), and 5.7±1.5% (4 to 7%), respectively (Su 2016). Results of reducing temperature test showed that there was no significant difference between the absorption freezer by using electricity and biogas combustion to power the absorption freezer. The temperature inside the absorption freezer was reduced to 0°C in 4 h by using electricity or biogas combustion as the energy to power the freezer.

The biogas absorption freezer was modified as the biogas absorption chiller by filling up with 150 L tap water and equipped with a set of water radiator as well as other related devices (Figure 7). Biogas was desulfurized and compressed for supplying steady biogas flow to the absorption chiller. Tap water was pumped into the absorption chiller and out of the chiller through a water radiator by a peristaltic pump. The sensors of temperature data logger were installed outside the freezer, inlet of tap water, inside the freezer under 150 L tap water, and outlet of the cool water (Figure 7). Results showed that the outlet water temperature was kept under 16 and 10°C under the water flow of 3 and 12 L/min, respectively. The temperature inside the chiller was kept below 6°C for both water flow sets (Figure 8). Inlet water temperature did not affect the water cooling significantly.

4.4 *Biogas upgrading*

Study of livestock biogas upgrading and biogas powered car was performed from 1974 in Taiwan by TLRI (Hong et al. 1980). For traditional automobile, a biogas/air mixer must be installed along with the carburetor to achieve biogas/air ratio of 1:7.5 and engine speed was

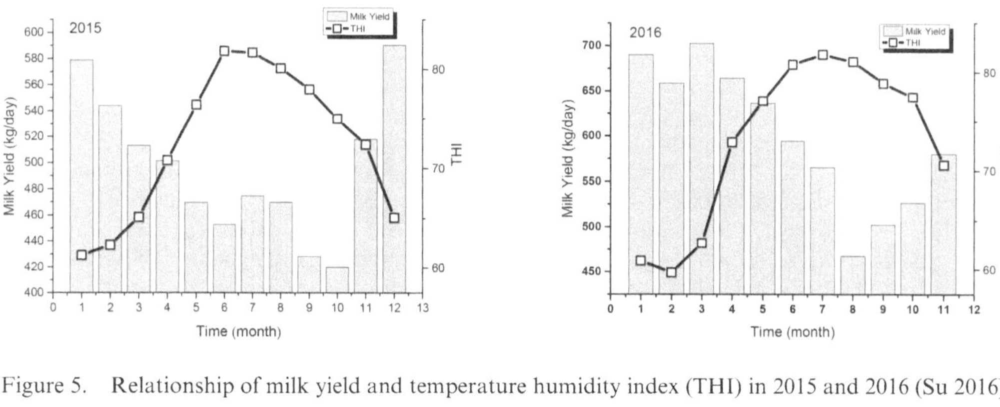

Figure 5. Relationship of milk yield and temperature humidity index (THI) in 2015 and 2016 (Su 2016).

Figure 6. Appearance of anaerobic digesters and biogas storage bag at the dairy farm of NTU.

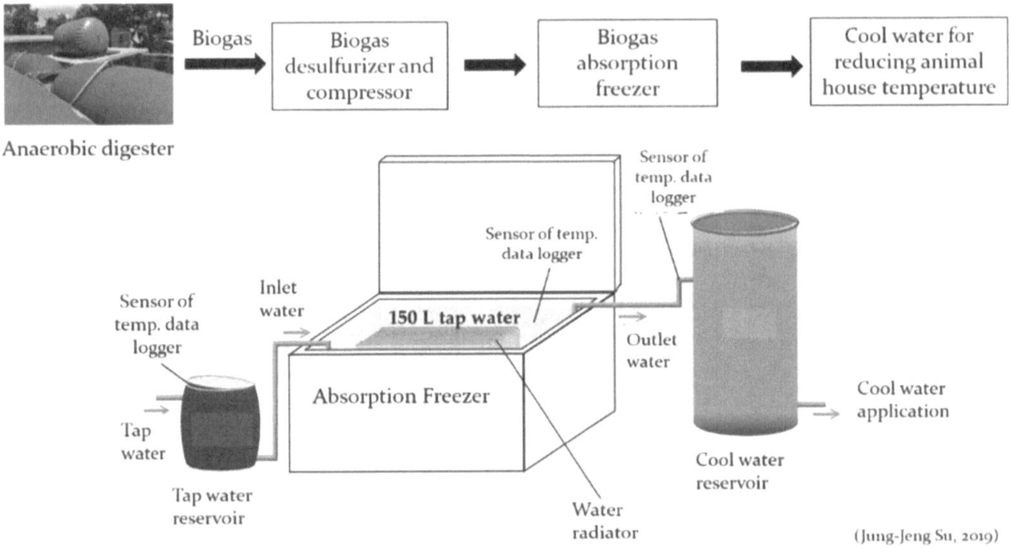

Figure 7. Sketch of biogas absorption chiller modified from absorption freezer for cool water application (Su 2016).

maintained around 700–6000 RPM for optimal operation. For example, a classic car with a displacement of 1,400 cc equipped with two 28-L gas cylinders of compressed biogas (150–200 kg/cm^2) could run for about 153 km/m^3 (biogas consumption rate = 73–83 km/m^3 biogas) (Chiu 1990). However, this technique was not studied continuously after 1990.

In 2019, the NTU/AST research team starts to develop a novel biogas upgrading technique by integrating domestic photocatalytic biogas desulfurizer and CO_2 absorption facility (Figure 9) (Su 2019). The pilot-scale photocatalytic is constituted of four acrylic cuboids in series with two Ultraviolet light tubes (120 cm length, 40 Watt UV fluorescent black light, PULSAR, China) attached to each acrylic cuboid (130 cm height × 20 cm width× 20 cm length, total volume = 52 L). Aluminum foil sheet was used to cover the surface area of the two acrylic cuboids for reflecting UV light towards the inside of the acrylic cuboids. The two acrylic cuboids were packed with the mixture of Rasching ring (i.e. hollow spherical polypropylene balls) (Sheng-Fa Plastics, Inc., Taiwan) and light-expanded clay aggregates (LECA) (Su et al. 2013). The pilot-scale domestic

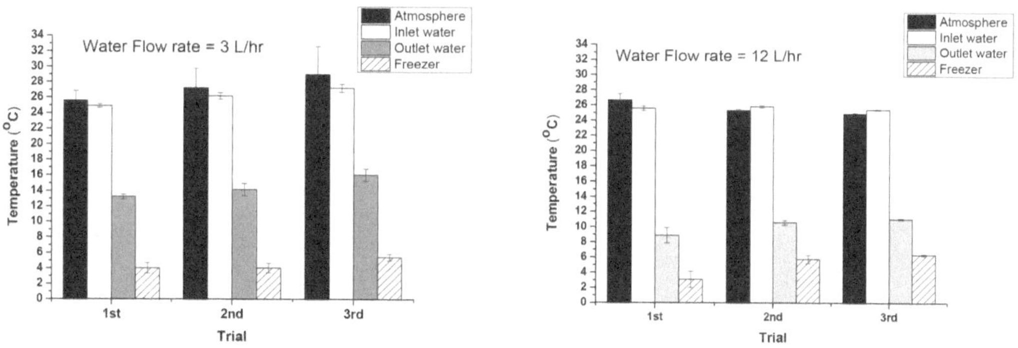

Figure 8. Changes of water temperature through a biogas absorption chiller under different inlet tap water flow rates (Su 2016).

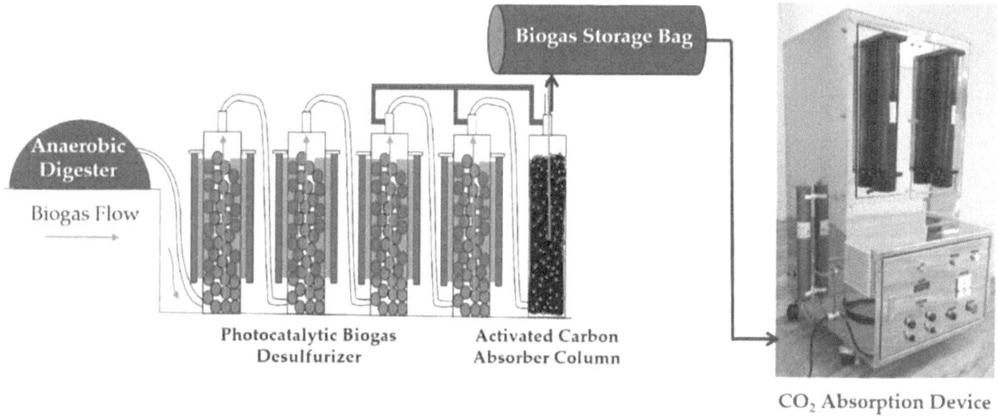

Figure 9. The sketch of biogas photocatalytic desulfurization and upgrading (Su 2019).

CO_2 absorption facility (AuraMat-HFDCO$_2$-CH4-1L-VCis, Aura Material Inc., Hsin-Chu, Taiwan) constituted of four hollow fiber filters, two filters in series, and alternative operation (Figure 9) (Su 2019).

This facility can be regenerated automatically inside the facility and the working pressure is 10 bar. The untreated raw biogas was introduced to the pilot-scale photocatalytic desulfurizers and then stored in a biogas bag (Figure 9).

Then, the desulfurized biogas was introduced to the pilot-scale CO_2 absorption facility after passing through an activated carbon filter column for drying and removing any impurity in the biogas. Results of preliminary study showed that methane concentration in all biogas samples after photocatalytic desulfurization was from 58.0±1.6 to 64.8±1.7% (Table 3). Removal efficiency of CO_2 in the biogas was 96.9±1.4% when the outlet biogas flow rate was 1 L/min (i.e. inlet biogas flow rate is about 2 L/min) (Table 4). The NTU/AST research team is planning to scale up the biogas upgrading facility to a 1,500-head commercial pig farm located in eastern Taiwan and equipped with a bio-desulfurization facility in the near future.

Table 3. The preliminary study of photocatalytic biogas desulfurizer (n = 30) (Su 2019).

Outlet biogas (L/min)	Inlet H_2S (ppm)	Outlet H_2S (ppm)	Removal (%)	CO_2 (%)	CH_4 (%)	N_2 (%)
0.5	3256±483	3.4±6.1	>99%	29.5±2.3	61.7±2.8	8.8±1.9
1	3733±462	14.8±25.7	>99%	30.0±0.6	62.0±2.3	8.0±2.3
1.5	3800±283	0	100%	31.5±0.9	58.0±1.6	10.6±2.6
3	3967±57	0.7±1.2	>99%	26.8±1.9	60.6±3.3	12.6±3.6
4	3950±100	3.65±7.3	>99%	27.2±1.8	64.8±1.7	7.98±2.9
5	3950±71	0	100%	27.8±1.4	64.7±0.6	7.5±0.9

Table 4. The preliminary study of CO_2 absorption facility after biogas desulfurization (n = 24) (Su 2019).

Outlet biogas L/min	CH_4 (%)			N_2 (%)			CO_2 (%)		
	Inlet	Outlet	Difference	Inlet	Outlet	Difference	Inlet	Outlet	Removal
1	64.4	85.5±1.1	+32.7	10.2	13.7±0.9	+34.6	25.4	0.8±0.4	96.9±1.4
1.5	62.1	76.3±3.0	+22.9	10.9	25.4±17.6	+10.0	27	10.0±4.2	62.9±15.4
2	61.3	72.8±4.6	+18.7	11.1	13.9±1.5	+25.1	27.6	13.2±5.4	52.1±19.6

4.5 *Biogas direct combustion*

Natural gas is the common fuel for gas cooker and stove in Taiwan. However, compressed and desulfurized livestock biogas can be applied as the fuel for household. Based on the early study of the TLRI, the pore size of the biogas injector of gas stove as well as hot water stove is enlarged to about 1.5 cm and the estimated biogas consumption is about 0.7 m^3/h. The volume of biogas consumption for hot water stove was 34, 114, and 197 L for the water temperature of 40, 60, and 90oC, respectively. Moreover, the volume of hot water production was 17.4, 7.0, and 2.9 L/min for the volume of biogas consumption of 34, 114, and 197 L, respectively (Chow et al. 1981).

In Taiwan, the energy structures of power generation by the Taiwan Power Company are gas combustion (30.1%), coal combustion (30.1%), private power generation plant (17.5%), nuclear power (9.3%), renewable energy (4.9%), fuel oil combustion (4.8%), combine heat and power (1.8%), and water power (1.4%) in 2017 (https://www.taipower.com.tw/tc/Chart.aspx?mid=194). Thus, thermal power generation is still the major power source in Taiwan. By the way, the main fuel for the thermal power generation plant includes natural gas (47.8%), coal (47.8%), heavy fuel oil (4.3%), and light fuel oil (0.2%) (https://www.taipower.com.tw/TC/page.aspx?mid=216). Thus, the desulfurized and upgraded livestock biogas has the potential to become bio-natural gas for substituting both coal and fuel oil in thermal power generation as well as natural gas for household use.

5 CONCLUSIONS

Biogas production from livestock solid waste and wastewater has been investigated. The most efficient way to reduce the greenhouse gas emission from livestock farm is to utilize the biogas completely (Shaw 2019, Krajewski 2019, Bartniczak et al. 2019). Some mitigation strategy of livestock greenhouse gas emission has been studied by means of biogas utilization. Biogas power generation is the main promotion policy by the Taiwanese Environmental Protection Administration (TEPA) and Council of Agriculture (COA) since 2016. The weather is warm or even hot (e.g. 30 to 38°C) in Taiwan and no public heating or hot water is needed during the whole year. Thus, the exhaust heat from biogas power generator normally is not collected for producing steam as well as electricity from steam turbine. Practically, cooling air or water from biogas combustion by an absorption chiller should be the best way of biogas applications in the tropical and subtropical areas such as Taiwan. Additionally, the desulfurized bio-natural gas should be an alternate fuel for thermal power generation in the near future.

ACKNOWLEDGEMENTS

The study was made possible by the grants awarded from the Council of Agriculture (COA), (Project No. 105AS-2.4.2-AD-U1; No. 105AS-6.2.1-AD-U2; 100AS-7.2.1-AD-U1 (2); 107AS-2.4.2-AD-U1(2); 108AS-17.2.1-AD-U3), and the Ministry of Science Technology (MOST) (Project No. MOST 105-2622-B-002-013 -CC2; MOST 106-2313-B-002-043), Executive Yuan, Taiwan.

REFERENCES

Baere, L.D. & Mattheeuws, B. 2010. Anaerobic digestion of MSW in Europe. BioCycle 51: 24–26.

Bartniczak, B. & Raszkowski A. 2019. Sustainable development in Asian countries - Indicator-based approach. *Problems of Sustainable Development/Problemy Ekorozwoju*. 14(1): 29–42.

Chiu, T.H. 1990. Biogas Utilization. Special Issue of Taiwan Livestock Research Institute (TLRI). No.10, Council of Agriculture (COA), Executive Yuan, Taiwan (In Chinese).

Chow, T.Y., Hong, C.M., Koh, M.T., Chiu, T.H. & Chung, P. 1981. Utilization of biogas in Taiwan. Proceedings of the Second International Symposium on Anaerobic Digestion, Germany.

COA. 2018. 2017 Capital food supplement, Agriculture Statistics Yearbook, Council of Agriculture (COA), Taiwan, ROC (http://agrstat.coa.gov.tw/sdweb/public/inquiry/InquireAdvance.aspx). (in Chinese).

COA. 2019. Pig Production Report (May 2019), Council of Agriculture (COA), Taiwan, ROC (http://agrstat.coa.gov.tw/sdweb/public/book/Book.aspx). (in Chinese).

de Laclos, H.F., Desbois, S. & Saint-Joly, C. 1997. Anaerobic digestion of municipal solid organic waste: Valorga full-scale plant in Tilburg, The Netherlands. *Water Sci. Technol.* 36: 457–462.

Hong, C.M., Koh, M.T. & Chow, T.Y. 1980. Production and utilization of biogas in Taiwan. Proceedings of International Symposium on Biogas, Microalgae & Livestock Wastes—Animal Waste Treatment and Utilization. pp. 117–123.

Hsiao, T.S., Chong, C.S., Gi, Y.J., Su, T.M. & Cheng, M.P. 2017. Brochure of biogas power generation demonstration in pig farm. Special Issue of Taiwan Livestock Research Institute (TLRI). No.168, Council of Agriculture (COA), Executive Yuan, Taiwan. (In Chinese).

IPCC. 2006. Chapter 10. Emissions from livestock and manure management. In: Dong, H., Mangino, J., McAllister, T.A., Hatfield, J.L., Johnson, D.E., Lasssey, K.R., de Lima, M.A. & Romanovskaya, A. (Eds.), 2006 IPCC Guidelines for National Greenhouse Gas Inventories, Vol. 4: Agriculture, Forestry and Other Land Use.

Kothari, R., Pandey, A.K., Kumar, S., Tyagi, V.V. & Tyagi, S.K. 2014. Different aspects of dry anaerobic digestion for bio-energy: An overview. *Renew. Sustain. Energy Rev.* 39: 174–195.

Krajewski, P. 2019. Bioeconomy – opportunities and dilemmas in the context of human rights protection and environmental resource management. *Problems of Sustainable Development/Problemy Ekorozwoju* 14(2):71–79.

Møller, H.B., Sommer, S.G. & Ahring, B.K. 2004. Methane productivity of manure, straw and solid fractions of manure. *Biomass Bioenerg.* 26: 485–495.

Pezzolla, D., Maria, F.D., Zadra, C., Massaccesi, L., Sordi, A. & Gigliotti, G. 2017. Optimization of solid-state anaerobic digestion through the percolate recirculation. Biomass Bioenerg. 96: 112–118.

Shaw, K. 2019, Implementing sustainability in global supply chain. *Problems of Sustainable Development/Problemy Ekorozwoju* 14(2): 117–127.

Su, J.J. 2016. Research on the feasibility of biogas combustion for cooling purpose on site. Final report of project no. 105AS-2.4.2-AD-U1. Council of Agriculture, Executive Yuan, Taiwan, ROC. (In Chinese).

Su, J.J. 2017. Development and demonstration of farm scale biogas bio-filter systems for livestock biogas applications in Taiwan (Chapter 2: Advances in Renewable Energy Research) (ISBN: 978-1-138-55367-5). London, UK: CRC Press. August, 2017: 7–24.

Su, J.J. 2019. Feasibility study of applying desulfurized livestock biogas to produce bio-natural gas. Mid-term report of project no. 108AS-17.2.1-AD-U3. Council of Agriculture, Executive Yuan, Taiwan, ROC. (In Chinese).

Su, J.J. & Chen, Y.J. 2018. Monitoring of greenhouse gas emissions from farm-scale anaerobic piggery waste-water digesters. *J. Agri. Sci.* 156(6): 739–747.

Su, J.J., Liu, B.Y. & Chang, Y.C. 2003. Emission of greenhouse gas from livestock waste and wastewater treatment in Taiwan. Agri. Ecosys. Environ. 95: 253–263.

Su, J.J., Liu, Y.L., Shu, F.J. & Wu, J.F. 1997. Treatment of piggery wastewater treatment by contact aeration treatment in coordination with the anaerobic fermentation of three-step piggery wastewater treatment (TPWT) process in Taiwan. *J. Environ. Sci. Heal.* 32A (1): 55–73.

TEPA. 2019. 2018 National Greenhouse Gas Inventory Report, Executive Summary. Taipei, Taiwan, ROC: Environmental Protection Administration, Executive Yuan Taiwan (TEPA) (http://unfccc.saveoursky.org.tw/2018nir/uploads/00_abstract_en.pdf).

Wee, C.Y. & Su, J.J. 2019. Biofuel produced from solid-state anaerobic digestion of dairy cattle manure in coordination with black soldier fly larvae decomposition. *Energies 12*: 911 (total 19 pages).

Xu, F.Q., Wang, Z.W., Tang, L. & Li, Y.B. 2014. A mass diffusion-based interpretation of the effect of total solids content on solid-state anaerobic digestion of cellulosic biomass. *Bioresource Technol.* 167: 178–185.

The Role of Agriculture in Climate Change Mitigation – Pawłowski, Litwińczuk & Zhou (eds)

Impacts of regeneration on soil respiration after clear-cutting in Chinese fir forests

Y. Wang & X. Zhu
Zhejiang A&F University, Hangzhou, China

1 INTRODUCTION

Chinese fir forests cover almost 11000000 ha in subtropical China, representing a key component of Chinese forest carbon sink because of their large area and fast growth (Wang et al. 2009). Forest soils as one of the most important carbon sink and source play significant roles in global warming, and soil respiration releases about 50-75 Pg C yr^{-1}, contributing the major flux in the global carbon cycle (Raich & Schlesinger 1992). Previous study indicated soil properties like soil temperature, soil moisture and root mass can affect soil respiration remarkably (Yashiro et al. 2008), while harvest and reforestation in plantation forests may change soil environment directly, thus potentially influence soil respiration and carbon stocks (Laporte et al. 2003). Clear-cutting is the predominant harvesting method (Ma et al. 2013), followed by either natural regeneration alone or combined with artificial regeneration in Chinese fir forests. However, how these practices impact soil respiration is not clear.

The decomposition of logging slash on the clear-cut site may compensate the decreases in root and rhizosphere respiration (Jandl et al. 2007). Due to changes in soil temperature (Kim 2008), soil moisture (Lavoie et al., 2013), soil water table depth (Zerva & Mencuccini 2005), root activity (Pangle & Seiler 2002), decomposition rate of soil organic matter (Binkley 1986), and soil pH (Kim 2008), clear-cutting affects soil respiration. Contradictory results were found in previous studies (Aguilos et al. 2014, Zerva & Mencuccini 2005, Gao et al. 2015, Yashiro et al. 2008), which may be attributed to differences in clear-cutting practices, vegetation (Laporte et al. 2003) and environmental factors (Jandl et al. 2007, Krajewski 2019, Shi et al. 2019, Bartniczak & Raszkowski, 2019).

Moreover, reforestation methods applied in various combinations to facilitate the growth and survival of new seedlings (Jandl et al. 2007), and may cause additional soil disturbance comparing to natural regeneration, further affect soil temperature, soil moisture and soil respiration.

Mounding may either increase soil respiration by mixing organic matter with soil (Giasson et al. 2006), or decrease soil respiration resulting from the remove of vegetation (Levy Booth et al. 2016). Fertilization can both increase (Jassal et al. 2010) and reduce soil respiration (Mojeremane et al. 2012). Due to the various combinations of reforestation practices, estimating the effect of regeneration on soil respiration is complicated.

The aim of this study was to compare the different impacts between natural regeneration and artificial regeneration in Chinese fir plantations on the environmental controls and soil respiration.

2 MATERIALS AND METHODS

The experiment was conducted from April 2014 to December 2015 in the township of Qingshan, Lin'an County, Zhejiang Province, China. The study area has subtropical monsoon climate with mean annual temperature of 16.4°C and a precipitation of 1629 mm (Wu et al. 2015). The weather condition during the study period showed in Figure 1. The Chinese fir forests in this study originated from reforestation in 1989 after a clear-cut. In 2014 the stand had

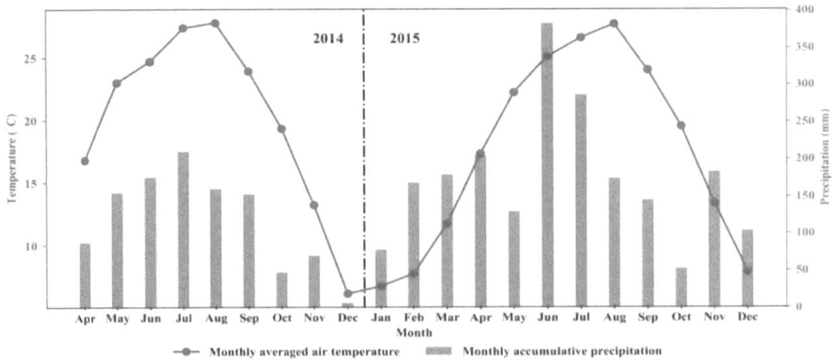

Figure 1. Mean average air temperature and monthly accumulative precipitation during the experimental period.

1350 stems ha-1 with a mean diameter at breast height of 16.5 cm, a mean height of 13.8 m, and a canopy closure of 88%. The shrub layer was dominated by *Mallotus apelta, Lindera glauca*, and *Eurya rubiginosa.*

In February 2014, 12 blocks (40 m×40 m) were established in a randomized block design with three replicates per treatment. The treatments were as follows: uncut control (CK), clear-cut with natural regeneration only (CN), and clear-cut followed by overall soil preparation, tending and replanting (CA). Main characteristics of treatment plots before and after clear-cutting showed in Table 1. Soil respiration was measured twice a month from April 2014 to December 2015 using the closed-chamber method (Yashiro et al. 2008). During the gas sampling, soil temperature at the depth of 5 cm was measured adjacent to each chamber. Soil moisture at 0-10 cm depth was measured gravimetrically by drying 10 g of soil at 105°C for 24 h.

To evaluate overall treatment effects on soil temperature, soil moisture content and soil respiration, repeated measures analysis of variance with least significant difference were applied using plot mean values in each month. To assess the effect of clear-cutting with reforestation on soil respiration, analysis of a nonlinear regression model was conducted with treatment category as the between-groups variable. Analyses were performed using IBM SPSS Statistics 23.0 (IBM Corp. Armonk NY) and Matlab 2016a. Results were considered statistically significant when $P < 0.05$.

3 RESULTS

Significant differences in soil temperature, soil moisture and soil respiration were detected between treatments by repeated-measures ANOVA ($P<0.05$). Mean soil temperature in clear-cut treatments was 1.0-1.8°C higher than CK ($P < 0.05$), following the order of CK < CU < CR (Figure 2a). Mean soil moisture content was generally in the order of CK > CU > CR during the entire duration of the experiment (Figure 2b). The mean moisture content in CU, CR was 8.01%,

Table 1. Main characteristics of treatment plots before and after clear-cutting.

	CK	CU	CA
Biomass of litter before cutting (Mg ha^{-1})	4.8±0.4a	5.0±0.7a	5.1±0.8a
Biomass of standing trees before cutting (Mg ha^{-1})	170.6±113a	165.1±90a	172.6±117a
Biomass of standing trees after cutting (Mg ha^{-1})	170.6 ±11.3a	0b	0b
Area of disturbed soil (%)	0a	2±0a	95±2b
Biomass of residues after clear-cutting (Mg ha^{-1})	0a	20.6±1.8b	21.5±1.7b
Biomass of litterfall and tending vegetation from Mar. 2014 to Dec. 2015 (Mg ha^{-1})	2.4±0.2a	0.0±0.0b	7.1±0.4d

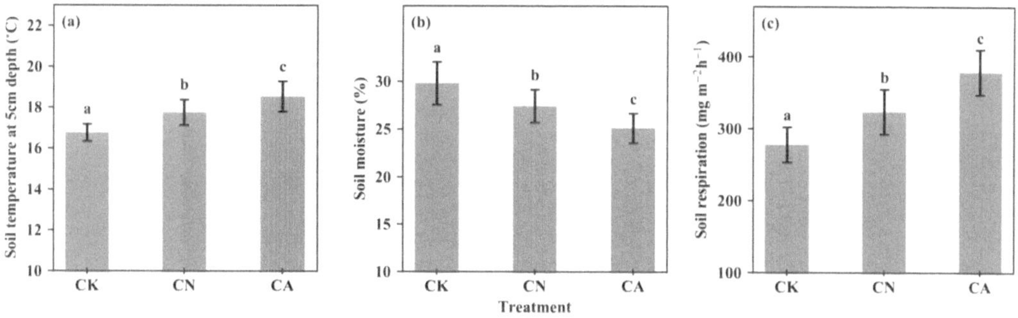

Figure 2. Mean soil temperature, soil moisture and soil respiration of each treatment during the experimental period, uncut control (CK), clear-cut with natural regeneration only (CN), and clear-cut followed by overall soil preparation, tending and replanting (CA). Different letters in the same column indicate statistically significant differences (n=3, $P < 0.05$).

and 15.76%, respectively, lower than that in CK. The mean soil respiration in the CU, CR was 16.50%, 36.38%, respectively, higher than that in CK ($P < 0.05$). The monthly dynamic of soil temperature, soil moisture and soil respiration in each treatment were showed in Table2.

4 DISCUSSION AND CONCLUSION

In our study, clear-cutting followed by regenerations caused a significant increase in the soil respiration. Clear-cutting alone increased respiration by 16.50% compared with CK. Soil respiration increase was slightly lower than in other studies that showed increases between 24.32% and 74.62% (Ullah et al. 2009, Kim 2008). Similar to our study, in all of these studies the harvest residues were left on the site, as was in our study. Soil preparation in this study where the residues were mixed throughout the seedling pits may expose seedling root systems to higher amounts of organic matter, thereby releasing nutrients in the rooting zone and enhancing the growth of seedlings. Compared with CU, the increased soil respiration in CA suggests that soil preparation and replanting result in extra soil respiration. The elevated soil respiration in the cut areas can be attributed to increased organic input, increase in disturbed soil surface area, and organic material decomposition rate (Stromgren et al. 2016). The decomposition of large amounts of logging slashes and root biomass of cut shrubs and herbs may compensate for the decrease in root and rhizosphere respiration compared to areas without soil treatment (Pumpanen 2003). In our study, the logging slash increased the amount of organic matter after clear-cutting. Compared to the CK, the amount of dead organic matter was on the average 3.6, 4.7 times higher in CU and CR, respectively. The residues represented an additional source of available C, which may partially explain the higher soil respiration. The physical disturbance of the soil by mixing and burying organic matter into the mineral soil in reforestation plots accelerate the rate of decomposition (Levy-Booth et al. 2016), and hence stimulate microbial activity and increase microbial abundance (Kuzyakov 2006). Soil preparation lead to exposure of the mineral soil and disruption of soil aggregates, which affected soil temperature and moisture. Overall soil preparation in CA resulted in decomposition and mineralization compared with conditions in CN with intact soil surface, which ultimately lead into differences in soil respiration.

In conclusion, there were significantly higher soil respiration after clear-cutting which increased with the intensity of disturbance. Soil preparation and tending could contribute with seedling survival and aid in the development of mixed forests, while causing higher soil respiration. From the perspective of carbon sequestration, natural regeneration is preferable than artificial regeneration for construct new stands after clear-cutting.

Table 2. Monthly value of soil temperature, soil moisture and soil respiration during the experiment period in each treatment. The values are mean ± SD. Different letters in the same row indicate statistically significant differences (n=3, $P < 0.05$).

		Temperature (℃)			Moisture (%)			Soil respiration ($mg \cdot m^{-2} \cdot h^{-1}$)		
Year	Month	CK	CN	CR	CK	CN	CR	CK	CN	CR
2014	Apr	14.1±0.29a	15.3±0.61b	15.9±0.67b	28.3±2.68a	27.3±1.09a	22.5±0.88b	195.0±13.40a	283.3±36.56ab	332.4±68.97b
	May	20.1±0.41a	21.4±0.59b	21.8±0.71b	35.7±3.55a	32.2±2.43a	28.5±1.10b	385.8±28.39a	465.8±40.95ab	531.3±46.30b
	Jun	22.6±0.74a	24.3±0.97b	24.8±1.10b	30.4±1.35a	26.9±1.12ab	25.8±1.48b	434.5±11.88a	502.9±25.44b	560.7±26.07c
	Jul	25.3±0.53a	25.7±1.16a	26.8±1.57c	40.7±2.47a	36.9±1.93b	36.2±1.31b	502.4±33.84a	567.1±60.26b	672.6±6.32c
	Aug	23.0±1.22a	23.2±1.81a	25.3±1.52c	35.6±1.02a	32.8±1.90a	31.4±2.02a	427.7±47.46a	453.4±55.47a	552.6±61.88b
	Sep	22.5±0.70a	23.0±1.08a	25.0±1.62c	28.9±0.45a	27.1±0.45a	25.0±1.18a	415.1±5.80a	440.7±20.33a	531.3±4.89b
	Oct	16.8±0.19a	17.4±0.65ab	18.3±1.03b	16.7±2.25a	15.1±2.07a	13.0±1.84a	271.7±25.98a	316.9±31.86b	358.4±27.89b
	Nov	12.8±0.05a	13.9±0.05b	14.5±0.40b	24.2±0.92a	20.7±0.97ab	18.4±1.09b	137.8±1.18a	151.4±13.32a	208.7±5.68b
	Dec	8.0±0.05a	8.1±0.16a	8.9±0.04b	8.5±1.19a	7.4±1.09a	6.8±1.06a	66.0±1.61a	74.1±3.70a	101.3±21.97b
2015	Jan	4.1±0.11a	4.1±0.06a	6.1±0.26b	23.3±1.78a	22.7±1.75a	22.0±1.65a	57.6±11.48a	75.7±4.22b	95.9±7.98c
	Feb	6.5±0.18a	6.5±0.14a	7.1±0.23b	32.9±1.88a	30.1±1.96b	26.9±1.25c	66.5±11.94a	83.6±21.02ab	104.6±14.39b
	Mar	12.4±0.36a	13.3±0.90a	13.6±0.68a	30.1±2.22a	28.3±3.33ab	26.2±1.89b	138.3±21.15a	180.2±45.05ab	213.8±16.78b
	Apr	14.6±0.31a	15.8±0.56b	16.5±0.82b	33.1±2.07a	32.1±0.91a	27.2±3.38b	192.8±33.96a	283.3±36.56b	332.4±68.97b
	May	19.9±0.42a	21.3±0.62b	21.7±0.63b	35.6±3.36a	32.1±2.09ab	28.2±1.74b	394.8±31.96a	458.3±51.49ab	513.6±55.35b
	Jun	22.8±0.61a	24.6±0.98ab	25.0±1.48b	38.9±1.25a	36.6±0.84ab	35.6±1.90b	434.0±32.75a	485.4±34.05ab	566.8±51.51b
	Jul	24.9±0.56a	27.1±0.61ab	27.7±0.42b	37.2±3.44a	34.8±1.83a	33.0±2.78a	534.0±28.43a	583.6±24.19a	642.3±18.16b
	Aug	23.8±0.35a	25.8±0.40b	26.0±1.75b	28.8±3.93a	27.7±1.28a	25.6±2.39a	451.9±12.79a	508.5±14.95a	582.6±37.96b
	Sep	20.6±0.51a	22.4±0.46b	23.0±0.40b	33.4±4.15a	30.6±1.32ab	28.3±0.29b	323.7±47.08a	383.4±19.71b	432.9±22.14b
	Oct	16.8±0.35a	18.4±0.40b	18.7±0.40b	29.7±4.16a	25.3±3.36ab	21.9±2.89b	219.9±52.84a	261.0±57.50a	286.6±48.63a
	Nov	11.3±0.59a	12.3±0.50b	12.6±0.45b	27.3±2.84a	25.1±2.86a	22.5±2.79a	121.1±29.09a	159.5±33.24a	216.7±32.51b
	Dec	8.9±0.21a	8.9±0.26a	9.6±0.29b	27.2±0.66a	24.5±1.72b	23.1±1.21b	58.1±27.89a	79.3±24.00ab	119.8±16.85b

REFERENCES

Aguilos, M., Takagi, K., Liang, N., Ueyama, M., Fukuzawa, K., Nomura, M., Sakai, R. 2014. Dynamics of ecosystem carbon balance recovering from a clear-cutting in a cool-temperate forest. *Agric. For. Meteorol.* 197:26–39.

Bartniczak, B. & Raszkowski, A. 2019. Sustainable Development in Asian countries – Indicator-based Approach. *Problemy Ekorozwoju/Problems of Sustainable Development* 14(1):29–42.

Binkley, D. 1986. Forest Nutrition Management. Wiley, New York.

Gao, S., Chen, J., Tang, Y., Xie, J., Zhang, R., Tang, J., Zhang, X.D. 2015. Ecosystem carbon (CO_2 and CH_4) fluxes of a Populus dettoides plantation in subtropical China during and post clear-cutting. *For. Ecol. Manage.* 357:206–219.

Giasson, M.A., Coursolle, C., Margolis, H.A. 2006. Ecosystem-level CO_2 fluxes from a boreal cutover in eastern Canada before and after scarification. *Agric. For. Meteorol.* 140:23–40.

Jandl, R., Linder, M., Vesterdal, L., Bauwens, B., Baritz, R., Hagedorn, F., Johnson, D.W., Minkkinen, K., Byrne, K.A. 2007. How strongly can forest management influence soil carbon sequestration. *Geoderma* 137:253–268.

Jassal, R.S., Black, T.A., Trofymow, J.A., Roy, R., Nesic, Z. 2010. Soil CO_2 and N_2O flux dynamics in a nitrogen-fertilized Pacific Northwest Douglas-fir stand. *Geoderma* 157:118–125.

Kim, C. 2008. Soil CO_2 efflux in clear-cut and uncut red pine (*Pinus densiflora* S. et Z.) stands in Korea. *For. Ecol. Manage.* 255(8):3318–3321.

Kuzyakov, Y. 2006. Sources of CO_2 efflux from soil and review of partitioning methods. *Soil Biol. Biochem.* 38(3):425–448.

Laporte, M.F., Duchesne, L.C., Morrison, I.K. 2003. Effect of clear-cutting, selection cutting, shelterwood cutting and microsites on soil surface CO_2 efflux in a tolerant hardwood ecosystem of northern Ontario. *For. Ecol. Manage.* 174:565–575.

Lavoie, M., Kellman, L., Risk, D. 2013. The effects of clear-cutting on soil CO_2, CH_4, and N_2O flux, storage and concentration in two Atlantic temperate forests in Nova Scotia, *Canada. For. Ecol. Manage.* 304:355–369.

Levy-Booth, D.J., Prescott, C.E., Christiansen, J.R., Grayston, S.J. 2016. Site preparation and fertilization of wet forests alter soil bacterial and fungal abundance, community profiles and CO_2 fluxes. *For. Ecol. Manage.* 375:159e171.

Krajewski, P. 2019. Bioeconomy – Opportunities and Dilemmas in the Context of Human Rights Protection and Environmental Resource Management. *Problemy Ekorozwoju/Problems of Sustainable Development* 14(2):71–79.

Mojeremane, W., Rees, R.M., Mencuccini, M. 2012. The effects of site preparation practices on carbon dioxide, methane and nitrous oxide fluxes from a peaty gley soil. *Forestry* 85:1–15.

Pangle, R.E., Seiler, J. 2002. Influence of seedling roots, environmental factors and soil characteristics on soil CO_2 efflux rates in a 2-year-old loblolly pine (*Pinus taeda L.*) plantation in the Virginia Piedmont. *Environ. Pollut.* 116:S85–S96.

Pumpanen, J. 2003. CO_2 efflux from Boreal Forest Soil before and after Clear-cutting and Site Preparation. University of Helsinki, Finland.

Raich, J.W., Schlesinger, W.H. 1992. The global carbon flux in soil respiration and its relationship to vegetation and climate. *Tellus* 44:81–99.

Shi, Y. B., Zhao, X. X., Jang, C-L., Fu, Q. Chang C-P. Looking at the Impacts of Income Inequality on Environmental Governance in China. *Problemy Ekorozwoju/Problems of Sustainable Development*, 14(2):63–70.

Stromgren, M., Hedwall, P.O., Olsson, B.A. 2016. Effects of stump harvest and site preparation on N_2O and CH_4 emissions from boreal forest soils after clear-cutting. *For. Ecol. Manage.* 371:15–22.

Ullah, S., Frasier, R., Pelletier, L., et al. 2009. Greenhouse gas fluxes from boreal forest soils during the snow-free period in Quebec. *Can. J. For. Res.* 39(3):666–680.

Wang, Q., Wang, S., Zhang, J. 2009. Assessing the effects of vegetation types on carbon storage fifteen years after reforestation on a Chinese fir site. For. Ecol. Manage. 258(7):1437–1441.

Wu, J.Q., Wang, Y.X., Yang, Y., Zhu, T.T., Zhu, X.D. 2015. Effects of crop tree release on stand growth and stand structure of Cunninghamia lanceolata plantation. *Chin. J. Appl. Ecol.* 26 (2):340–348.

Yashiro, Y., Kadir, W.R., Okuda, T., Okuda, T., Koizumi, H. 2008. The effects of logging on soil greenhouse gas (CO_2, CH_4, N_2O) flux in a tropical rain forest, Peninsular Malaysia. *Agric. For. Meteorol.* 148(5):799–806.

Zerva, A., Mencuccini, M. 2005. Short-term effects of clearfelling on soil CO_2, CH_4 and N_2O fluxes in a Sitka spruce plantation. *Soil Biol. Biochem.* 37:2025–2036.

The Role of Agriculture in Climate Change Mitigation – Pawłowski, Litwińczuk & Zhou (eds)
 ISBN 978-0-367-43372-7

Impact of waste and by-products management on soil carbon sequestration

M. Pawłowska, M. Chomczyńska & M. Zdeb
Faculty of Environmental Engineering, Lublin University of Technology, Lublin, Poland

G. Żukowska & M. Myszura
Institute of Soil Science and Environment Management, Faculty of Agrobioengineering, University of Life Science in Lublin, Lublin, Poland

1 INTRODUCTION

New solutions in waste management targeted at improving the efficiency of waste recovery enable to incorporate the residue materials into the natural biogeochemical cycles. Land spreading, consisting in: introducing the waste into the soil to increase the fertility or improve the quality (use as fertilizer or soil improver) as well as restoring biological activity of degraded soil (use in land reclamation), plays a special role in this phenomenon. The organic waste and by-products, such as digestates deserve a special mention in this group. These residues are considered not only as a source of easily available nutrients for soil microorganisms and plants, but also as a source of organic matter, that enriches the soil with humic substances (HS). Thus, they become the source of exogenous organic matter (EOM) which is defined as all organic material of biological origin applied to the soil in order to fertilize, amend or restore it and improve the environment (Marmo et al. 2004). It is widely recognized that the soil organic matter (SOM) determines numerous soil properties. Therefore, its quantity and quality affect the directions of chemical and biochemical transformations occurring in the soil, influencing the development of soil microbial communities, their diversity, structure and activities (Bonilla et al. 2012).

According to the data given by Mondini et al. (2018) about 1,200 mln tons of EOMs are produced in Europe each year (excluding the biomass of crop residues). Assuming that the utilized agricultural area in Europe is 178,700 ml ha (Eurostat, 2016) and the average soil bulk density is 1.3 m^3/ton, incorporation of the total amount of EOMs into the 0.3 m thick layer of the soil covering the whole agricultural area would increase the SOM content by about 0.2%. Naturally, it does not mean that the entire pool of carbon contained in the organic matter can be retained in soils. A large part will be released into the atmosphere as a result of the mineralization process. However, even a small increase of SOM can significantly improve the poor conditions of the European soils. It is assessed that ca. 45% of the mineral soils in Europe have low or very low organic carbon content (in the range of 0 to 2 %) (Louwagie et al. 2009), and that the soils in southern Europe have a significantly lesser carbon content (Zdruli et al. 2004).

Therefore, land spreading represents an effective option for sustainable management of organic waste contributing to the enhancement of the soil organic carbon (SOC) content. Such a way of waste application falls within the scope of the activities recommended by the European Commission specified in Thematic Strategy for Soil Protection (COM(2006)231)), which was adopted by the European Parliament in a Resolution on the Thematic Strategy for Soil Protection (OJ WE C282 E/141). The decrease in the content of organic matter in soils is mentioned in this document as one of the main threats to the soil quality. This threat should be countered by a widespread the use of stable organic matter (OM) that is contained in compost, manure, and in much lesser extent, in sewage sludge and animal slurry. This stable fraction of OM contributes to the increase in humus pool in the soil (TSSP 2008).The importance of typological diversity of soils, which requires an individual approach to solving problems on a local scale, is emphasized in the document. The Strategy draws attention to the strong relationship between soil and climate change, pointing to their interactions and highlighting the impact of changes in soil use on the climate conditions. The changes can be beneficial, contributing to the alleviation of greenhouse effect through the C sequestration, or on the contrary, leading to an increase of greenhouse gases emission.

The soil provides substantial support in C sequestration. It is estimated that soil contains 2500 Gt of C that is ca. 80% of total C stored in all terrestrial ecosystems (Lal 2008). Over 60% of soil C is bound in organic compounds (Lal 2004). The soil carbon pool is approximately 3 times higher than the atmospheric one (Oelkers & Cole 2008). The potential of C binding in soils is not entirely known. It depends on many factors, such as soil properties (especially grain size distribution and water content) and climate conditions (Ontl & Schulte 2012). The modifications in soil properties can work in two directions, leading to a decrease or increase in the organic C pool of terrestrial ecosystems.

Enhancement of soil C sequestration should be taken at least as seriously as the reduction of the anthropogenic CO_2 emission. The world's attention should be directed to the problem of soil humus depletion, and the methods of counteracting this phenomenon should be indicated. It seems that the rational use of C resources contained in the waste may mitigate this problem.

1.1 *Relationships between waste management and carbon sequestration*

Waste management can have a direct and indirect impact on the C sequestration in terrestrial ecosystems. Direct impact is associated with the introduction of stable organic matter with organic waste into the surface layer of the lithosphere. Carbon is partially integrated into humic substances (HS) that are composed of slowly biodegradable, hardly soluble or insoluble in soil water solution compounds with high molecular weight (Table 1) modified during mineralization and humification processes from the plant and animal tissues. Humic substances fulfil multiple functions in soil formation and its fertility. These compounds govern the sorption properties, as well as the water and air conditions; therefore, they affect most of the soil transformations. Embedding C into some types of HS is a long-term process. Decomposition of HS in soil may take from 15 to over hundreds of years, depending on the environment conditions (Prusty & Azzez 2005). Stockmann et al. (2013) defined the soil C sequestration as long-term (i.e. >100 years) or permanent removal of CO_2 from the atmosphere and its "lock up" into the soil. Some part of the C introduced into the soil along with organic waste is assimilated by heterotrophic soil microorganisms as a result of their metabolic processes.

Table 1. Selected properties of humic substances.

Property	Fulvic acids	Humic acids	Humins	Source
Chemical formula	$C_{135}H_{182}O_{95}N_5S_2$	$C_{187}H_{186}O_{89}N_9S$	Not known	(Prusty & Azeez, 2005)
Molecular weight	1,000 to 10,000	10,000 to 100,000	100,000 to 10,000,000	Pettit, 2002
Colour	Yellow	Brown or grey	Black	Stevenson, 1982
Solubility	Soluble in water at all pH conditions	Soluble in water under alkali conditions, insoluble under acid condition	Insoluble in acid and alkali	Pettit, 2002
Susceptibility to degradation	Most susceptible to microbial attack	Intermediate resistance to microbial attack	Most resistant to microbial attack	(Prusty & Azeez, 2005)
Soil residence time	15-50 years	Over 100 years	Not known	(Prusty & Azeez, 2005)

Consequently, carbon is temporarily bound in soil microorganisms. Part of that C will be released from soil due to mineralization, but the remaining part will be incorporated into humus.

Another link between waste management and C bio-sequestration relates to the recycling of paper and cardboard. Such activities enable to reduce harvesting of the trees, used as a primary source of cellulose. By recycling the paper waste, the C retention in the forest biomass will be increased (Ackerman 2010). However, the results of the analysis obtained by Tatoutchoup (2016) showed that only increasing the rate of paper recycling to the certain level (considered as optimal) will result in expanding the surface area of the forests. However, setting too high recycling goals (beyond the optimal level) will result in decreasing the forested area.

1.2 *Legal approach to land spreading of the waste*

In Poland, the admission of waste to be recovered in form of land treatment for agricultural benefit or ecological improvement is specified in the Regulation of 20 January 2015 on the R10 recovery process that constitutes the executive law of the Waste Act of 14 December 2012 on waste. These regulations refer to the European Union law contained in the Directive on Waste (2008/98/EC).

The annex to this regulation lists the types of waste allowed for such recovery, and it describes the conditions for their use. The following residues are authorized for land application: stabilized sewage sludge from municipal wastewater treatment, sewage sludge from the treatment of certain types of industrial wastewater, e.g. fruit and vegetable processing, waste of cereal and animal products, waste from sugar, dairy and bakery industries, waste from alcohol production, waste plant biomass (properly ground and prepared using biological methods), livestock manure, sawdust, shavings, wood cuttings, forestry waste, straw middlings and bran from the plant feed production, waste bark, fruit pomace, pulp, compost produced from green waste and other bio-waste that does not meet the quality requirements, digested waste after dry and wet fermentation of biodegradable waste collected separately, biodegradable landscaping-generated wastes, liquids and other waste from anaerobic decomposition of animal and vegetable waste, post-treatment therapeutic peat mud. Individual types of waste must meet the conditions specified in the relevant regulations which guarantee maintaining the good environmental status. The waste must not pose a chemical and sanitary threat to any of the elements of the environment.

1.3 *Selected waste and by-products tested as a source of soil organic matter*

1.3.1 *Biochar*

Biochar is "a solid material obtained from the thermo-chemical conversion of biomass in an oxygen-limited environment" (IBI 2012). Biochar is produced from different types of lignocellulosic biomass, also the waste-derived one. Usually, the term "biochar" refers to the material that is intended for use in the non-oxidative applications, e.g. as the soil amendments, while the product used as a fuel is called "charcoal" (Hagemann et al. 2018).

Soil application of biochar has a long term story. Biochar has been used in agricultural practices in the Amazon Basin of South America in the ancient times (Cunha et al. 2009). It is uncertain whether terra preta do indio (black soil of the Indian) was created intentionally. However, this question is irrelevant. The most important is that the long-term effects of biochar application to the soils can be observed. The content of organic matter in terra preta is up to 150 g kg^{-1} (Petersen et al. 2001), which is few times higher than in typical soils in the surroundings. Glaser et al. (2001) stated that key factor in the persistence of organic matter in these soils is black carbon which is the residue of incomplete biomass combustion. Black C can exist in the environment for hundreds to thousands of years (Haberstroh et al. 2006).

The unique potential of C sequestration related to the biochar results from high stability of this element. The carbon contained in biochar is mainly integrated into the aromatic structures (Lehman & Joseph 2009) which makes it stable. The assessment of biochar stability is the subject of many studies. It is made on the basis of the analysis of biochar C structure, biochar oxidation resistance, biochar persistence on mineralization, aromaticity and degree of aromatic condensation, as well as the thermal recalcitrance index (Leng et al. 2019). One of the stability criteria is also O:C ratio. In the case of biochar, its value is below 0.3, and indicates a high degree of aromaticity (Baldock & Smernick 2002). Bolan et al. (2012) stated that the decomposition rate measured by half-life ($t_{1/2}$) value for green waste biochar (9989 days) was 53 times higher than the value of $t_{1/2}$ for poultry manure compost. Spokes (2010), based on the literature studies, stated that the $t_{1/2}$ for biochar is correlated to the O:C molar ratio. When the ratio is lower than 0.2, biochar is usually the most stable and its $t_{1/2}$ value is estimated as more than 1000 years, when the O:C ratio is between 0.2 and 0.6, the intermediate stability is estimated as $t_{1/2}$ 100-1000 years; in turn, and when the O:C ratio exceeds 0.6, biochar is the least stabile, with $t_{1/2}$ reaching 100 years.

The results of the study conducted by Smith (2016) indicate that biochar has useful negative CO_2 emission potential (~0.7 GtCeq. yr^{-1}). Additionally, biochar could be implemented in combination with the other negative emissions technologies, i.e. bioenergy with C capture and storage. Lehmann et al. (2006) estimated that up to 12% of the anthropogenic emissions of C caused by land-use change (0.21 Pg C) could be eliminated by applying C to the soil in the form of carbonized biomass. However, taking into account the CO_2 emission from fossil fuel combustion, the applications of biochar together with enhanced silicate weathering probably will not able to balance more than 5% of annual emissions from this source (Schlesinger & Amundson 2019).

Increasing interest in the conversion of organic residues into biochar and its incorporation into the soils in order to mitigate climate changes has been observed recently. Cayuela et al. (2010) examined the dynamics of C transformation in sandy soil amended with ten by-products of bioenergy production processes, i.e. pyrolysis, anaerobic digestion, transesterification. The laboratory batch tests showed that biochars were the most stable residues, which showed the lowest CO_2 emissions. The C loss determined after 60 days of incubation under 20°C ranged from 0.5% to 5.8% of total C added. Additionally, among all the tested residues, only biochar decreased the emission of N_2O – the other important GHG gas. This phenomenon can be explained by high specific surface area, large amount of chemically reactive sites and high porosity of the biochar (Zhang et al. 2011). A similar study was conducted by Galvez et al. (2012). They examined four bioenergy by-products: digestate, rapeseed meal, distiller grains, biochar, and commonly used organic amendments: sewage sludge and two kinds of composts. Among these amendments only the biochar did not cause any significant changes in soil respiration, N availability and enzymatic activity. These observations are not consistent with common opinion that high porosity, specific surface area and electrochemical charges of the biochar significantly modify the soil system, changing surface area, pore-size distribution and affecting the physical and physicochemical properties of the soil (Downie et al. 2012). This inconsistency can be explained by a short period of study (30 days) conducted by Galvez et al. (2012), which was not sufficient to obtain a significant change in the soil microbiology. However, it was observed that biochar promoted the C accumulation; thus, the researchers concluded that biochar significantly favors C sequestration, although it does not improve the soil fertility. Significant influence of biochar on C retention in soil, both as alone additive and in combination with digestate was observed by Cardelli et al. (2018) in short-term (100-days) laboratory tests.

1.4 *Aerobically stabilized organic materials*

Compost and different organic residues can be also used to stabilize C in soils (Bolan et al. 2012). There are two main types of compost applications: as an additive used for improving soil properties or fertilizing, and as a component of growing media. Compost is considered a multifunctional soil fertilizer. It supplies macro- and microelements, although the supply of plant-available nitrogen by compost is usually insufficient (Kluge et al. 2008). Compost contains HS that are a crucial indicator of its maturation; thus, the compost addition to the soil leads to an increase in the pool of stable organic matter.

The long-term field study of Albiach et al. (2001) conducted on sandy-silty loam has confirmed the possibility of C sequestration in soil by its amendment with compost. The researchers stated that MSW compost used in dose 24 t $ha^{-1}yr^{-1}$ yielded the highest increases in the contents of OM, total HS, and humic acids, compared to sewage sludge and ovine manure used in the same dose. However, short-term laboratory study conducted by Robin et al. (2018) showed less stability of the OM applied to the sandy loam soil with green waste compost compared to dairy manure and vermicompost. The differences in the results obtained by the afore-mentioned authors can be explained by many factors related to both the characteristics of the compost itself, e.g. its stability, but also the duration of the tests, external conditions and even the characteristics of the soil itself. Bolan et al. (2012) stated that the clay content in a soil significantly decreases the rate of decomposition of organic substances, enhancing the C stabilization in soil, which can be attributed to C immobilization through preventing its microbial decomposition. In their study, the value of half-life ($t_{1/2}$) for decomposition of OM contained in the poultry manure compost ranged from 139 days in the sandy soil and 187 days in the clay soil. Additionally, they observed that the addition of clay materials, such as gibbsite, goethite, and allophane to the compost applied into sandy soil slowed down the OM decomposition rate, increasing the value of $t_{1/2}$ from 139 (compost without additives) to 806, 620, and 474 days, respectively. The clays application did not deteriorate the utility properties of compost for soil improving (Bolan et al. 2012). The explanation for the relationship between C stabilization and clay particles content may be the results of studies conducted by Lehmann et al. (2007) that pointed at a strong role for organo-mineral interactions in C stabilization within soil micro-aggregates. They observed a positive correlation between the amount of O–H groups on kaolinite surfaces and aliphatic C forms (r^2 in the range of 0.66–0.75), which suggests the crucial role of aliphatic C-H groups in organo-mineral interactions. Finally they concluded that the adsorption of organic compounds on mineral surfaces is initially the dominant process of C stabilization in soil.

1.5 *Anaerobically stabilized waste/digestates*

Digestate is a residue of anaerobic digestion (AD) process. It consists of undigested substances, products of feedstock decomposition, cells of microorganisms, and water. Digestates from AD plant are generally used as fertilizers, because of a high content of the available N, P and K fractions. This residue contains ca. 90% of volatile solids (Monlau et al. 2016); thus, it contributes to soil C turnover.

Field and laboratory studies showed that the organic C contained in digestates can be retained in soils. Béghin-Tanneau et al. (2019) estimated that the use of digested maize silage favored C sequestration and reduced the CO_2 emissions by 27% as compared to the use of undigested maize silage. Cayuela et al. (2010) observed that only 40% of the C stock added to the soil with manure digestate had been emitted in the form of CO_2, in a laboratory batch study. Digestate showed a higher stability compared to the nonfermentables from hydrolysis of lignocellulosic biomass (60% of C added was emitted), and compared to rapeseed meal and distillers grains (80% of C added was emitted). Only the C derived from biochar was incomparably more stable (less than 5.8% of C added was emitted). More significant share of pyrochar (biochar obtained during pyrolysis) in C sequestration, compared to solid-digestate was also observed by Monlau et al. (2016). They stated that these products showed complementary effects in soil: pyrochar sequestered C, and digestate enhanced microbiological processes in the soil.

The diversity of the results obtained by different authors on C sequestration in the soil amended with digestates can be explained by the stability level. Alburquerque et al. (2012) showed in the short-term soil incubation tests conducted on six digestates obtained during co-digestion of pig and cattle slurry as the main substrates, that the C sequestration significantly depends on the degree of digestate stabilization. The unstable digestates used in their study emitted significant amounts of CO_2.

1.6 *Obstacles in estimation of soil carbon sequestration potential*

Despite clear evidence on the possibility of increasing C stocks in soils by enriching them in EOM from waste and by-products, the question arises: to what level can the concentration of stable C in soil be increased. The answer to this question is the precondition for estimation of soil C sequestration potential, both on a local and a global scale.

There is a balance in natural ecosystems between the synthesis of the new substances and their degradation (Bonilla et al. 2012). It is commonly known that the addition of harvested residues to the soil is a way to increase the SOM content in croplands. However, the input of the OM is not linearly related to the stock of SOM. Thus, increasing the dose of organic residues added to the soil will cause the increase of SOM content only to the certain level (Shahbaz et al. 2017). This level determines the soil saturation with carbon.

Generally, the soils in Europe are most likely to be accumulating carbon (EEA 2012) which suggests that they have not met the C saturation level. This statement is particularly true in the case of the soils under grassland and forests that are considered as a C sink (estimated up to 80 million tons of C per year), whereas the soils under arable land are a C source, emitting 10–40 million tons of C per year (EEA 2012). These soils exceeded their ability to retain the C under certain conditions. They contribute to net CO_2 emissions. Stewart et al. (2007) emphasize the importance of the C saturation level for the soil capacity to C sequestration. Their model study based on the results of long-term field experiments showed that the greatest efficiency in soil C sequestration was observed in the soils that were further from the C saturation value.

The factors that limit the saturation level and rate of C sequestration in soil are not yet sufficiently understood, but it is nevertheless known that the transformation of EOM into HS results from the physical, chemical and biological soil properties, climatic conditions, land use and management (Franzluebbers 2004), as well as the properties of EOM source, e.g. its stability (that was mentioned in sec. 3). Shahbaz et al. (2017) stated on the results of short-term (64 days) study that the occlusion of OM supplied with wheat residues within soil aggregates depended on the EOM dose and type of biomass (leaves, stalks, and roots). An increase in the residue dose from 1.40 to 5.04 g DM kg^{-1} stimulated aggregate formation, but the proportion of residues occluded within aggregates decreased with the increasing dose. Mineralization of biomass enhanced with the EOM dose, but it was lesser for roots than for the aboveground parts of the plant.

Dependence of the C accumulation in soil on such many factors causes that the effects of EOM application on SOC stocks are very difficult to predict. The variability of soil and EOM composition and properties, spatial variability of SOC and the significant differences in environmental conditions influence the C turnover in soils (Mondini et al. 2018). Additionally, transformation of SOC in soil is usually very slow. For example, in temperate regions, significant changes in the SOC content were not observed even 10–20 years after a considerable modification in land management (Franzluebbers 2004). Such a late environmental response also makes the estimation of C sequestration difficult and unreliable. However, many researchers attempt to evaluate the C sequestration possibilities in soils. For example, Mondini et al. (2018) calculated that rate of potential C sequestration in Italian soils is in the range 0.110–0.385 t C ha^{-1} y^{-1} depending on EOM types. They used the RothC model, modified by Mondini et al. (2017), and they assumed the continuous amendment (for 100 years) of soils with EOM under the climate change conditions. These values are lower than the rates of potential C sequestration estimated by Freibauer et al. (2004) and Smith et al. (2008) for the amendment European soils that are 0.40 t C ha^{-1} y^{-1} and 0.42 t C ha^{-1} y^{-1}, respectively.

2 CONCLUSIONS

Waste and by-products are a huge reservoir of organic matter, which, when introduced into the soil, could contribute to increasing the soil C stock. However, the long-term retention of C introduced into the soil is problematic. The results of the literature review on soil C sequestration show that C pool in the soil does not grow proportionally to the increase of the amount of C introduced into the soil along with EOM, and the efficiency of C sequestration in soil depends on many factors related to the properties of soil and EOM source, climatic conditions, soil depth, to which the organic matter were introduced, etc. The analysis of the results of research on the fate of organic matter derived from various types of waste applied to the soil indicates that the greatest prospects for long-term C sequestration are associated with the application of biochar. However, other organic materials can also help to increase the pool of the soil carbon, especially since the biochar alone does not provide enough nutrients, which can slow down the rate of humification and plant growth processes, and consequently, the biosequestration. This points to the need to focus attention on seeking suitable multi-substrate mixtures that will contain highly stabilized forms of organic matter. The effect of anaerobiosis on the biochar-derived C sequestration in soil is also worth of interest, because it is a poorly understood issue.

The review showed another important field for further research. It is an analysis of the levels of C saturation in soil, and determination of the factors that affect them. Understanding this issue is crucial for reliable estimation of the soil potential for C sequestration on local and global scales.

REFERENCES

Ackerman F. 2000. Waste management and climate change. Local environment. *The International Journal of Justice and Sustainability* 5(2): 223–229.

Act on Waste of 14 December 2012 (Journal of Laws of 2013, item 21).

Albiach R., Canet R., Pomares F., Ingelmo F. 2001. Organic matter components and aggregate stability after the application of different amendments to a horticultural soil. *Biores. Technol.* 76(2): 125–129.

Alburquerque J.A., de la Fuente C., Bernal M.P. 2012. Chemical properties of anaerobic digestates affecting C and N dynamics in amended soils. *Agriculture, Ecosystems & Environment* 160: 15–22.

Baldock J. A., Smernik R. J. 2002. Chemical composition and bioavailability of thermally altered *Pinus resinosa* (Red pine) wood. *Organic Geochemistry* 33: 1093–1109.

Béghin-Tanneau R., Guérin F., Guiresse M., Kleiber D., Scheiner J.D. 2019. Carbon sequestration in soil amended with anaerobic digested matter. *Soil and Tillage Research* 192: 87–94.

Bolan N.S., Kunhikrishnan A., Choppala G.K., Thangarajan R., Chung J.W. 2012. Stabilization of carbon in composts and biochars in relation to carbon sequestration and soil fertility. *Sci Total Environ.* 424: 264–70.

Bonilla N., Gutiérrez-Barranquero J.A. De Vicente A., Cazorla F.M. Enhancing. 2012. Soil Quality and Plant Health Through Suppressive Organic Amendment. *Diversity* 4(4): 475–491.

Cardelli R., Giussani G.A., Marchini F., Aaviozzi A. 2018. Short-term effects on soil of biogas digestate, biochar and their combinations. *Soil Research* 56(6): 623–631.

Cayuela M.L., Oenema O., Kuikman P.J., Bakker R. R., Van Groenigen J.W. 2010. Bioenergy by-products as soil amendments? Implications for carbon sequestration and greenhouse gas emissions. *GCB Bioenergy* 2(4): 201–213.

Cunha T.J.F, Madari B.E., Canellas L.P., Ribeiro L.P., de Melo Benites V., de Araújo Santos G. 2009. Soil organic matter and fertility of anthropogenic dark earths (Terra Preta de Índio) in the Brazilian Amazon basin. *Rev. Bras. Ciênc. Solo* 33(1): 85–93.

Directive 2008/98/EC of the European Parliament and of the Council of 19 November 2008 on waste and repealing certain Directives, Official Journal of the European Union L 312/3, 22. 11.2008.

Downie A., Crosky A., Munroe P. Physical Properties of Biochar. In: *Biochar for Environmental Management Science and Technology* (Eds. J. Lehmann, S. Joseph). Taylor&Francis Group, 2012.

EEA 2012. *Soil organic carbon*. European Environmental Agency (https://www.eea.europa.eu/data-and-maps/indicators/soil-organic-carbon-1/assessment.

European Parliament Resolution of 13 November 2007 on the Thematic Strategy for Soil Protection (OJ WE C 282 E from 6 November 2008): 138–144.

Eurostat. 2016 (https://ec.europa.eu/eurostat/web/products-datasets/-/tag00025).

Franzluebbers A.J. 2004. Organic residues, decomposition. In: *Encyclopedia of Soils in the Environment* (eds. D. Hillel, J.L. Hatfield, D. S. Powlson, C. Rosenzweig, K.M. Scow, M.J. Singer, and D.L. Sparks). Amsterdam: Elsevier/Academic Press: 112–118.

Galvez A., Sinicco T., Cayuela M.L., Mingorance M.D., Fornasier F., Mondini C. 2012. Short term effects of bioenergy by-products on soil C and N dynamics, nutrient availability and biochemical properties. *Agriculture, Ecosystems & Environment* 160: 3–14.

Glaser B., Haumaier L., Guggenberger G., Zech W. 2001. The 'Terra Preta phenomenon: a model for sustainable agriculture in the humid tropics. *Naturwissenschaften*. 88 (1): 37–41.
Haberstroh PR, Brandes JA, Gélinas Y, Dickens AF, Wirick S, Cody G. 2006. Chemical composition of the graphitic black carbon fraction in riverine and marine sediments at sub-micron scales using carbon x-ray spectromicroscopy. *Geochimica et Cosmochimica Acta* 70(6): 1483–1494.
Hagemann N., Spokas K., Schidt H-P, Kägi R., Böhler M.A. Bucheli T.D. 2018. Activated Carbon, Biochar and Charcoal: Linkages and Synergies across Pyrogenic Carnon's ABCs. *Water* 10(2): 182.
IBI. 2012. *Standardized Product Definition and Product Testing Guidelines for Biochar That Is Used in Soi*l (aka IBI Biochar Standards) Version 2.1 BI.
Lal R. 2004. Soil carbon sequestration impact on global climate change and food security. Science 304: 1623–1627.
Lal R. 2008. Carbon sequestration. *Philosophical Transactions of the Royal Society* B 363: 815–830.
Lehmann J., Gaunt J., Rondon M. 2006. Bio-char Sequestration in Terrestrial Ecosystems – A Review. *Mitig Adapt Strat Glob* Change 11: 403.
Lehmann J., Kinyangi J., Solomon D. 2007. Organic matter stabilization in soil microaggregates: implications from spatial heterogeneity of organic carbon contents and carbon forms. *Biogeochem.* 85(1): 45–57.
Leng L. Huang H., Li H., Li J., Zhou W. 2019. Biochar stability assessment methods: A review. *Sci. Total Environ.* 647(10): 210–222.
Louwagie G., Gay S.H, Burrell A. *Final report on the project 'Sustainable Agriculture and Soil Conservation (SoCo)*' JRC Scientific and Technical Reports Luxembourg: European Commission, JRC 2009.
Marmo L., Feix I., Bourmeau E., Amlinger F., Bannick C.G., De Neve S., et al. 2004. *Reports of the Technical Working Groups Established Under the Thematic Strategy for Soil Protection.* Volume-III. Organic Matter and Biodiversity. EUR 21319 EN/3. Luxembourg.
Mondini C., Cayuela M.L., Sinicco T., Fornasier F., Galvez A., Sánchez-Monedero M.A. 2018. Soil C Storage Potential of Exogenous Organic Matter at Regional Level (Italy) Under Climate Change Simulated by RothC Model Modified for Amended Soils. *Front. Environ. Sci.* 29 November 2018.
Mondini C., Cayuela M.L., Sinicco T., Fornasier F., Galvez A., Sánchez-Monedero M.A. 2017. Modification of the RothC model to simulate soil C mineralization of exogenous organic matter. *Biogeosciences* 14: 3253–3274.
Monlau F., Francavilla M. Sambusiti C., Antoniou N., Solhy A., Libutti A. et al. 2016. Toward a functional integration of anaerobic digestion and pyrolysis for a sustainable resource management. Comparison between solid-digestate and its derived pyrochar as soil amendment. *Appl. Energ* 169: 652–62.
Ning C., Gao P., Wang B., Linn W., Jiang N., Cai K. 2017. J. of Integrative Agriculture 16(8): 1819–31.
Oelkers E.H., Cole D.R. 2008. Carbon dioxide sequestration: a solution to the global problem. *Elements* 4: 305–310.
Ontl T.A., Schulte, L.A. 2012. Soil Carbon Storage. *Nature Education Knowledge* 3(10): 35.
Petersen J.B., Neves E.G., Heckenberger M.J. 2001. *Gift from the past: Terra preta and prehistoric amerindian occupation in Amazonia.* In: Unknown Amazonia (Ed. C McEwan). London, UK: 86–105.
Pettit RE. 2002. *Organic Matter, Humus, Humate, Humic Acid, Fulvic Acid and Humin: Their Importance in Soil Fertility and Plant Health*, Texas: A & M University: 1–24.
Powlson D.S., Riche A.B., Coleman K., Glendining M.J, Whitmore A.P. 2008. Carbon sequestration in European soils through straw incorporation: Limitations and alternatives. *Waste Manage.* 28(4): 741–6.
Prusty B.A.K., Azeez P.A. 2005. Humus: The Natural Organic Matter in the Soil System. *J. Agril. Res. & Dev.* 1: 1–12.
Regulation of 20 January 2015 on the R10 recovery process (Journal of Laws of 2015, item 132).
Robin P., Morel C., Vial F. 2018. Effect of three types of Exogenous Organic Carbon on Soil Organic Matter. *Sustainability* 10(4): 1146.
Schlesinger W.H., Amundson R. 2019. Managing for soil carbon sequestration: Let's get realistic. *Glob Chang Biol.* 25(2): 386–389.
Shahbaz M., Kuzyakov Y., Heitkamp F. 2017. Decrease of soil organic matter stabilization with increasing inputs: Mechanisms and controls *Geoderma* 304: 76–82.
Smith P. 2016. Soil carbon sequestration and biochar as negative emission technologies. *Global Change Biol.* 22(3).
Spokas K.A. 2010. Review of the stability of biochar in soils: predictability of O:C molar ratios. *Carbon Management* 1(2): 289–303.
Stevenson F.J. *Humus Chemistry*. Wiley, New York, 1982.

Stewart C.E., Paustian K., Conant R.T., Plante A.F., Six J. 2007. Soil carbon saturation: concept, evidence and evaluation. *Biogeochemistry* 86(1): 19–31.
Stockmann U., Adams M.A., Crawford J.W., Field D.J., Henakaarchchi N., Jenkins M., et al. 2013. The knowns, known unknowns and unknowns of sequestration of soil organic carbon. *Agric. Ecosyst. Environ.* 164: 80–99.
Tatoutchoup F.D. 2016. Optimal rate of paper recycling. *Forest Policy and Economics* 73: 264–269.
TSSP. 2008. *European Parliament Resolution of 13 November 2007 on the Thematic Strategy for Soil Protection* (2006/2293(INI)) (OJ WE C 282 E from 6 November 2008: 138–144).
Zdruli P., Jones R.J.A., Montanarella L. *Organic Matter in the Soils of Southern Europe European Soil.* Bureau Technical Report. Luxembourg: Publications Office of the European Union. 2004.
Zhang A., Liu Y., Pan G., Hussain Q., Li L., Zheng J., Zhang X. 2011. Effect of biochar amendment on maize yield and greenhouse gas emissions from a soil organic carbon poor calcareous loamy soil from Central China Plain. *Plant and Soil.* 351: 263–275.

The Role of Agriculture in Climate Change Mitigation – Pawłowski, Litwińczuk & Zhou (eds)
 ISBN 978-0-367-43372-7

Mitigation of climate changes by carbon sequestration in biomass under elevated atmospheric CO_2 concentration – facts and expectations

M. Chomczyńska, M. Pawłowska & A. Słomka
Faculty of Environmental Engineering, Lublin University of Technology, Lublin, Poland

G. Żukowska
Institute of Soil Science and Environment Management, Faculty of Agrobioengineering, University of Life Science in Lublin, Lublin, Poland

1 INTRODUCTION

Natural transformations of carbon compounds in the environment favor its retention in terrestrial and aquatic ecosystems, while the human activities mostly lead to the accumulation of C in the atmosphere, in the form of CO_2. The problem with anthropogenic emissions concerns mainly the imbalance of the CO_2 release and assimilation rates by terrestrial and aquatic ecosystems.

The anthropogenic emission influences the upper part of natural carbon cycle that occurs between the atmosphere, oceans, and terrestrial ecosystems, but it does not involve the carbon exchange with the geologic reservoirs. The exchange with the upper part of the carbon cycle occurs on the timescales ranging from sub-daily to millennia, while the exchange with the lower part needs the longer timescales (Archer et al. 2009). Therefore, there the return of a huge pool of coal to the zones from which they were released as a result of human activities, is impossible within a reasonable time scale.

The imbalance related to the rates of CO_2 assimilation and release processes leads to an increase of the atmospheric C accumulation. The value of this parameter rose from 3.3 ± 0.1 GtC yr^{-1} in 1980s (IPCC 2001), by 4.1 ± 0.1 GtC yr^{-1} in 2000-2005 (IPCC 2007) to 4.7 ± 0.1 GtC yr^{-1} for the years 2007-2016 (Le Quéré et al. 2018). The anthropogenic emissions resulting from industrial development are considered as the main reason of this phenomenon. Until the start of the industrial revolution in the 18th century, the global CO_2 emissions were at the level of 0.011 Gt CO_2 yr^{-1}. From 1760 onward, it began to slowly increase, reaching a 25-fold higher value after 100 years. A significant increase in CO_2 emissions occurred after the Second World War. By the 1960s, emissions had increased about 28 times compared to the value of 1860. During the following years, emissions grew further, reaching a short-time stabilization of 35.69 $GtCO_2$ yr^{-1} in 2014 and 2015. Unfortunately, the growth was observed again in 2016-2018 (Figure 1).

An increase in the CO_2 emissions from anthropogenic sources is related to the increase in the atmospheric CO_2 concertation (Figure 1). Since the pre-industrial era, when the CO_2 concentration was 280 ppm, it has increased by 1.5 times compared to the present time. An evident acceleration of CO_2 accumulation in the atmosphere has been observed since the 1950s. According to the data from Earth System Research Laboratory (ESRL), the recent monthly mean CO_2 concentration measured at Mauna Loa Observatory (Hawaii) in August 2019 was 408.53 ppm. This value is similar to the average CO_2 concentration measured for 2018. However, this is the lowest month value noted in 2019, and the others ranged from 408.54 to 414.66 ppm.

This continuous increase of the CO_2 concentration in the atmosphere, reaching 2-3 ppm per year (ESRL date), on average, indicates that the initiatives undertaken in recent years to reduce the CO_2 emissions are still insufficient. It is necessary to intensify two-way activities, not only aimed at reducing emissions, but also at increasing the rate of CO_2 removal from the

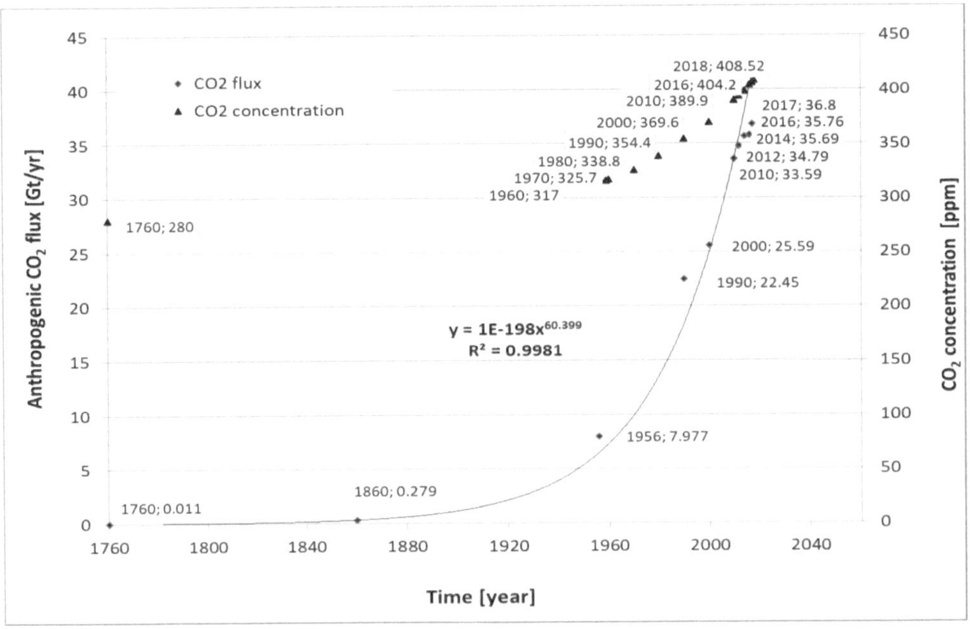

Figure 1. Time-depended changes of the anthropogenic CO_2 emission (based on the data of Le Quéré et al., 2018) and the atmospheric CO_2 concentration (ESRL data).

atmosphere. For the latter, the carbon sequestration in terrestrial ecosystems, including biomass, which is a natural carbon sink, is a prospective solution. It could be expected that due to the intensification of photosynthesis induced by the increase in substrate concentration, i.e. CO_2, biomass growth will be stimulated, which will lead to increased CO_2 absorption (Cel et al. 2016). As a result, the alleviation of atmospheric carbon increase will be achieved. However, this phenomenon is not so obvious, due to a number of factors, such as an increase in temperature due to the greenhouse effect, changes in evapotranspiration, caused by the physiological changes in plants and an increase in temperature, etc. However, the extremely important is the human activity. Further destruction of natural ecosystems, such as tropical forest or savannas may nullify the positive effects of the stimulating activity of the environmental factors, even if they occur.

The authors analyzed the results obtained during the examination of different plant species under different conditions, regarding the impact of increased CO_2 concentration in the atmosphere on the accumulation of carbon in biomass, and on the other phenomena that may affect the plant growth. The purpose of this review was to assess the reality of expectations for self-accelerating mitigation of greenhouse effect by enhancement of photosynthesis due to the increased availability of CO_2.

1.1 *Role of plants in global carbon cycle*

Carbon dioxide is a common carbon compound that is formed as a result of its oxidation, which can occur in various parts of the Earth's ecosystem, wherever carbon exists, and conditions are favorable. Carbon is present in all sub-systems of the global ecosystem. The Earth's crust (including soil) contains ca. 100 mln GtC in minerals and fossil fuels (the latter contain ca. 3,500 GtC); the oceans contain ca. 38,000 GtC in the water, surface sediments and marine biota; the atmosphere contains ca. 750 GtC, mainly in the form of CO_2, whereas the terrestrial biosphere contains 560 GtC in organic compounds that build plants and animals, as well as human tissues (IPCC 2007; Ajani et al. 2013), but the highest amount is fixed in tree biomass.

Soil, which is the surface layer of the Earth's crust, stands out from among other layers with biological activity, is an important carbon pool. It contains about 2,500 GtC primarily in the form of organic matter, derived from dead plants and microorganisms (Ajani et al. 2013).

In the Earth's ecosystem, carbon circulates between the individual subsystems. The most important process is the exchange of carbon between plants and the atmosphere.

Plants absorb CO_2 from the atmosphere, building the carbon into the organic matter during the photosynthesis. On the other hand, the respiration process releases CO_2 into the atmosphere. It is estimated that globally 450 Gt CO_2 yr^{-1} (i.e. approximately 123 Gt C yr^{-1}) is absorbed from the atmosphere due to photosynthesis (IPCC 2007). It is the largest stream of carbon flux in the entire global cycle of this element. On the other hand, 435 Gt CO2 yr^{-1} (i.e. approx. 119 Gt C yr^{-1}) is released into the atmosphere in the processes of plant and soil respiration, as well as fires (IPCC 2007). Thus, about 4 Gt C yr^{-1} is immobilized in the biomass as a result. Plant biomass is therefore a sink for carbon, despite the fact that the part of it contained in dead tissues undergoes mineralization, as a result of which carbon is released into the atmosphere. However, not all the carbon is released. Part of it is embedded into the resistant structures of humic compounds, which is considered as a form of sequestration because they can remain in the soil for hundreds of years (Prusty & Azzez 2005).

1.2 *Direct response of plant on elevated CO_2 concentration in the atmosphere*

Taking into account the mechanism of photosynthesis, the positive effects of increased CO_2 concentration in the atmosphere on plant productivity can be expected (Allen 1991). The literature review done by Kimball et al. (2002) confirms the reality of these expectations (Table 1). Moreover, the results of other studies carried out on different plant species showed the stimulating effect of the CO_2 concentration on growth of biomass yield. For example, Mortensen (1995) observed a 10% increase in mean relative growth rate of *Betula pubescens* Ehrh. seedlings caused by the increase of the CO_2 concentration in the atmosphere from 350 to 560 μmol mol^{-1}, during the 35-day greenhouse experiment. An increase in the biomass production was also observed by Jongen and Jonsen (1998) in the experiment carried out on the following grasses: *Lolium perenne* L., *Cynosurus cristatus* L., *Holcus lanatus* L. and *Agrostis capillarys* L. The plants were cultivated at the CO_2 concentration of 700 μmol mol^{-1}, in the 8-month experiment with the open-top chamber. The study showed that the examined grass species varied in their response to elevated CO_2 concentration in terms of the biomass growth. Wang et al. (2015) noted that the elevated CO_2 concentration (up to 600-699 ppm) enhanced the average rice yield by 20% as compared to the atmospheric conditions, but no significant increase in the grain size was observed. The response of rice and the examined grasses on the CO_2 concentration can be explained by their photosynthesis pathway. Generally, in terms of metabolic pathways for CO_2 fixation, the plants are divided into three groups: C_3, C_4 and CAM. The first and most numerous group includes the plants (including rice and most of grasses species) that use the Photosynthetic Carbon Reduction (PCR) also called Calvin-Benson cycle in which the enzyme: ribulose-1,5-bisphosphate carboxylase/oxygenase (RuBisCo) catalyzes the first step producing a three-carbon compound, phosphoglycerate (3-PGA). This enzyme catalyzes two competing reactions: carboxylation and oxygenation (Portis & Parry 2007). The oxygenation leads to loss the fixed carbon ranging between 25% and 30%. Therefore, minimizing the oxidizing potential of RuBisCo may increase carbon assimilation (Long et al. 2006). The carbon fixation pathways used by C_4 and CAM plants have additional steps that enable reducing the loss of carbon during the process. The C4 photosynthesis (Hatch-Slack pathway) is aimed at limiting the photorespiration by the spatial separation of CO_2 fixation and carbon assimilation. It is possible because of different structure of mesophyll and bundle sheath cells (Lara & Andreo 2011). The process is named for the 4-carbon intermediates (malic acid or aspartic acid) that are produced in the first stage in mesophyll cells via the conversion of CO_2 to phosphoenolpyruvate (PEP). The particles of the intermediates are transported to the bundle sheath cells where decarboxylation process occurs and CO_2 and C_3 organic acids are produced. These acids return to the mesophyll cells to regenerate PEP (Ludwig 2016). The concentration of CO_2 in the bundle

Table 1. Effect of increased CO_2 level on photosynthesis of different plants.

Plant	Photo-synthesis pathway	CO_2 level[a] [μmol mol^{-1}]	Increase due to growth of CO_2concentration compared to ambient CO_2 concentration [%]	References
Kentucky bluegrass (*Poa pratensis*), photosynthesis rate	C_3	+350	141	He et al. 1992
Big blustem (*Andropogon gerardi* Vitman), photo-synthesis rate	C_4	+350	no significant changes	He et al. 1992
Cotton (upper leaf) (*Gos-sypium hirsutum* L.), net photosynthesis	C_3	550	28.2	Hileman et al. 1994[b]
Wheat (upper leaf) (*Triti-cum aestivum* L.), net photosynthesis	C_3	550	31.5	Garcia et al. 1998[b]
Wheat (flag leaf) (*Triticum aestivum* L.), net photosynthesis	C_3	550	25.6	Osborne et al. 1998[b]
Ryegrass (7 day cut) (*Lolium perenne* L.), net photosynthesis	C_3	600	32.5	Rogers et al.1998[b]
Wheat (canopy) (*Triticum aestivum* L.), net photosynthesis	C_3	+220	19.2	Brooks et al. 2001[b]
Sorghum (upper leaf) (*Sorghum bicolor* L.), net photosynthesis	C_4	+200	8.5	Wall et al. 2001[b]
Rice (total biomass) (*Oryza sativa*) different cultivars net photosynthesis	C_3	600-699	20	Wang et al. 2015
Kyasuwa grass (total bio-mass), *Cenchrus pedicella-tus*, net photosynthesis	C_4	950	no significant changes	Tom-Dery et al. 2018

[a] in the studies a constant target CO_2 set point (CO_2 concentrations without "+") or a target increment in concentration above normal air CO_2 level was used (CO_2 concentrations preceded by "+")
[b] according to the review of Kimball et al. (2002); data obtained under free air CO_2 enrichment (FACE) conditions

sheath cells is 10-20 times higher than in mesophyll cells (Hatch 1999). In the C_4 plants, the photorespiration is inhibited by an increase of the CO_2 concentration that suppresses the oxygenase activity of RuBiSco inside the bundle sheath cells (Lara & Andreo 2011). The CAM (Crassulacean acid metabolism) pathway also involves a fixation of CO_2 into C_4 acids prior to the carbon fixation via RuBisCo. In CAM plants, the process of carbon fixation and assimilation are separated in time.

This fixation process occurs in night when the stomata in the leaves are opened and PEP carboxylase is active. During the day, stomata are closed to reduce transpiration, and then C_4 acids are decarboxylated, releasing the CO_2, which is fixed in normal C_3 photosynthesis pathway (Sage 2008).

Leakey et al. (2009) stated that the stimulating effect of the increase in the CO_2 concentration in ambient air is evident for the plants with C_3 photosynthetic pathway. In the case of these plants, the growth of the atmospheric CO_2 content raises the intercellular CO_2 concentration that accelerates the carboxylation of ribulose-1,5-bisphosphate – RubP (catalyzed by ribulose bisphosphate carboxylase-oxygenase) and suppresses photorespiration, which decreases the carbon release. The described mechanism is absent in the C_4 plants, in which the carbon uptake is saturated by carbon dioxide at lower intercellular CO_2 concentration achieved at current atmospheric CO_2 (the primary carboxylase of C_4 plants – phoshoenolopyruvate carboxylase has a lower Km for CO_2) and carbon dioxide is concentrated in bundle sheath cells that saturate the carboxylation of RubP. Under elevated CO_2 concentration in the atmosphere, the C_4 plants improve their water use efficiency reducing the stomatal conductance that indirectly stimulates photosynthesis in the periods or the places with drought stress (Leakey 2009).

The data given in Table 1 show that the intensification of photosynthesis due to the increased atmospheric CO_2 concentration can be considered as a factor contributing to the mitigation of climate changes. However, particular plant species react differently to the elevated concentration of CO_2 (Table 1, Table 2), and the growth of plants depends not only on the availability of CO_2.

Zheng et al. (2018) showed that grasses, such as tall fescue (Festuca arundinacea), perennial ryegrass (Lolium perenne) and Kentucky bluegrass (Poa pratensis) have the saturation value of atmospheric CO_2 concentration for their growth in the range between 400 to 1200 μmol mol^{-1}. Over this value, the net photosynthesis was not increasing. The fescue was characterized by the lower optimal value as compared to the other examined species. The study of Teixeira Da Silva et al. (2006) showed that the apparent optimal CO_2 concentration value significantly differs for particular plant species. They did not observe the optimum value for ornamental Spathiphyllum cv. even at 3000 μmol mol^{-1}. In their study, the total biomass grew of 61%, 127% and 185% at the CO_2 concentration of 1000, 2000 and 3000 μmol mol^{-1}, respectively, compared to the value obtained at 375 μmol mol^{-1}.

Table 2. The effect of increased CO_2 level on shoot and root biomass of different plants under free air CO_2 enrichment (FACE) conditions (Kimball et al. 2002).

Plant	Photo-synthesis pathway	CO_2 level[a] [μmol mol-1] shoot	root	Increase due to growth of CO_2 concentration compared to ambient CO_2 concentration [%] shoot	root	References
Wheat (*Triticum aestivum* L.)	C_3	+200	550	4.8 ÷ 11.7	27.9	Wechsung et al. 1999 Kimball et al. 2002
Rice (*Oryza sativa* L.)	C_3	589	589	7.6 ÷ 10.8	18.4 ÷19.5	Kim et al. 2001 Kimball et al. 2002
Ryegrass (*Lolium perenne* L.)	C_3	600	600	10.5 ÷ 20.2	29.3 ÷161.3	Daepp et al. 2000,2001, Hebeisen et al. 1997
Sorghum (*Sorghum bicolor* L.)	C_4	+200	-	-1.0 ÷ 6.7	-	Ottoman et al. 2001
Clover (*Trifolium repens* L.)	C_3	600	600	7.6 ÷ 38.3	5.6 ÷ 39.6	Hebeisen et al. 1997, Van Kessel et al. 2000
Potato (*Solanum tuberosum* L.)	C_3	560	-	-12.2 ÷ -28.6	-	Bindi et al. 1998, 1999
Cotton (*Gossypium hirsutum* L.)	C_3	550	550	32.3÷36.8	37.8 ÷ 156.9 (tap root) 29.3 ÷ 100 (fine roots)	Mauney et al. 1992, 1994, Prior et al. 1994, Rogers et al. 1992
Grape (*Vitis vinifera* L.)	C_3	550,700	-	20.6 ÷ 42.1	-	Bindi et al. 1995, 2001

[a] Explanation is the same as given under Table 1.

1.3 *Factors influencing plant growth under elevated atmospheric CO_2 concentration*

Numerous other factors, such as macro- and microelement contents in soil, water accessibility and temperature can influence the plant growth under the elevated CO_2 concentration in the atmosphere. Improper value of only one of them can limit the biomass increase even if the CO_2 concentration will be still rising.

This phenomenon is illustrated by the data presented in Table 3, which show the influence of the soil nitrogen concentration on the biomass growth examined under elevated CO_2 concentration. The data show that a lower increase in shoot biomass due to an increase of the CO_2 concentration was observed in the case of the soils with low nitrogen level compared to the soil with ample content of this macroelement. Sometimes, higher CO_2 concentration did not compensate the N deficiency, and then the plant biomass was lower as compared to that obtained on the soil with ample nitrogen level. Is such case, the nitrogen is the limiting factor.

It is worth considering here the impact of others factors on the plant biomass productivity (and at the same time on the biomass primary productivity – Table 4) the changes of which are also induced by increasing the CO_2 levels in the atmosphere. The carbon dioxide emission contributes to climate warming, increasing the air temperature. Experimental data and model studies show that the stimulation of plant productivity by higher CO_2 level increase with rising temperatures, although this regularity applied to the plants growing in non-stress temperature ranges (Allen (1991; Luo et al. 2008; Bernacchi et al. 2006). Therefore, it cannot be ruled out that if the critical temperature values are exceeded for some physiological phenomena in plants, productivity will be reduced despite the increased CO_2 concentrations (Leakey et al. 2012).

A temperature increase caused by elevated atmospheric CO_2 can reduce the moisture availability for plants, intensifying the evaporation process. On the other hand, at increased CO_2 plants exhibit lower water requirements (by reducing transpiration), which improves their water use efficiency (Parry 1990). According to Leakey at al. (2012) reduced water use at increased CO_2 levels enhances the plant productivity when they ameliorate water stress under the drought conditions – unfortunately at some level of drought stress, the biomass production can be limited to such extent that the impact of higher CO_2 concentration will not

Table 3. The effect of increased CO_2 level on shoot biomass of plants at different nitrogen level under FACE conditions (Kimball et al. 2002).

Plant	CO_2 level[a] [µmol mol^{-1}]	Increase due to growth of CO_2 concentration compared to the ambient CO_2 concentration [%]		References
		Ample N	Low N	
Ryegrass (*Lolium perenne* L.)	600	5.8	3.7	Hebeisen et al. 1997
Ryegrass (*Lolium perenne* L.)	600	8.0	-8.6	Hebeisen et al. 1997
Ryegrass (*Lolium perenne* L.)	600	10.9	1.6	Hebeisen et al. 1997
Ryegrass (*Lolium perenne* L.)	600	18.6	-6.2	Daepp et al. 2000
Ryegrass (*Lolium perenne* L.)	600	10.5	-0.3	Daepp et al. 2000
Rice (*Oryza sativa* L.)	589	10.8	8.1	Kim et al. 2001
Wheat (*Triticum aestivum* L.)	+200	11.7	2.8	Kimball et al. 2002
Ryegrass (*Lolium perenne* L.)	600	20.1	6.9	Daepp et al. 2000

[a] Explanation is the same as under Table 1

Table 4. The estimated effect of increased CO_2 level (~550 ppm) on net primary productivity (or seed yield) and photosynthetic CO_2 uptake of ecosystems (Leakey et al. 2012).

Biome	Primary functional groups	Change in: net primary productivity/seed yield [%]	photosynthetic CO_2 uptake) [%]
Temperate forest	C_3 trees	23±2a)	46±4c)
Temperate grassland	Community	13±4b)	37±7c)
	C_3 grasses		-2±9c)
	C_4 grasses		
Temperate cropland	C_3 crops	17±6b)	13±5c)
	C_4 crops	6±9b)	11±6c)

a) Estimate based on four FACE experiments from Norby et al. (2005)
b) Estimate based on updated meta-analysis of FACE data following methods of Ainsworth and Long (2005)
c) Estimates from Ainsworth and Rogers (2007)

compensate for the water deficiency. To summarize, it should be emphasized that the CO_2 fertilization effect in natural ecosystems is expected (Table 4) but its magnitude and duration is dependent on environmental factors, among which soil moisture, nutrients availability and temperature seem to be the most important (Leakey et al. 2012; Sneed 2018). Since the mentioned factors vary regionally, the response of biomes productivity to the elevated CO_2 is expected to be diverse (Rosenthal & Tomeo 2013). Accurately determining the impact of elevated carbon dioxide levels on the productivity of natural ecosystems requires further research, in particular regarding rainforests and tropical savannas that have a large share in the global carbon cycle (Leakey et al. 2012). Modelling studies indicate that tropical forest will respond more significantly to the elevated CO_2 concentration than temperate and boreal forests (Cernusak et al. 2013).

2 CONCLUSIONS

Terrestrial ecosystems, especially multi-species forests, are a strong absorber of the atmospheric carbon. Undoubtedly, an increasing the area covered by vegetation must increase the retention of this element. However, the human activity aimed at expanding the surface area of arable land, intense droughts and forest fires do not bode well. About 3.6 million ha of primary tropical forests, the area comparable with the surface area of Belgium, disappeared in 2018 (GFW). In such a situation, the expectation that the self-accelerating process of biomass growth due to increased CO_2 concentration will be a sufficient way to stop the process of C accumulation in the atmosphere does seem rational. The research shows that not all the plant species react to the increased CO_2 concentration by raising their yield. Type of the photosynthesis pathway is the crucial factor. Additionally, the plant response depends on environmental variables, such as soil nutrients abundance, water availability, and temperature. Taking these into account it can be expected that the enhancing accumulation of C in biomass due to the increase of concentration of CO_2 in the atmosphere will take place mainly in the case of the plants belonging to C3 photosynthesis group, growing on the fertile soils. Low soil richness in nutrients, especially phosphorus and nitrogen, seems to be a serious barrier in this process.

ACKNOWLEDGEMENTS

This work has been prepared as a part of the implementation of statutory tasks No. FN-75/IŚ/2019 of the Faculty of Environmental Engineering of Lublin University of Technology.

REFERENCES

Ainsworth E.A., Long S.P. 2005. What have we learned from 15 years of free air CO_2 enrichment (FACE)? A meta-analytic review of the responses of photosynthesis, canopy properties and plant production to rising CO_2. New Phytol, 165, 351–372.

Ainsworth E.A., Rogers A. 2007. The response of photosynthesis and stomatal conductance to rising [CO_2]: mechanisms and environmental interactions. Plant Cell Environ, 30, 258–270.

Ajani J.I., Keith H., Blakers M. G. Mackey B.G., King H.P. 2013. Comprehensive carbon stock and flow accounting: A national framework to support climate change mitigation policy. Ecological Economics 89: 61–72.

Allen L.H. Jr. Effects of Increasing Carbon Dioxide Levels and Climate Change on Plant Growth, Evapotranspiration, and Water Resources. In. Managing Water Resources in the West Under Conditions of Climate Uncertainty: A Proceedings. 1991. 101–147. (https://www.nap.edu/read/1911/chapter/8).

Archer D., Eby M., Brovkin V., Ridgwell A., Cao L., Mikolajewicz U., Caldeira K. M., K., Munhoven G., Montenegro A., Tokos K.: Atmospheric Lifetime of Fossil Fuel Carbon Dioxide, Annu. Rev. Earth Pl. Sc., 37, 117–134, 200.

Bernacchi C.J., Leakey A.D.B., Heady L.E., Morgan P.B., Dohleman F.G., McGrath J.M., Gillespie K.M., Wittig V.E., Rogers A., Long S.P., Ort D.R. 2006. Hourly and seasonal variation in photosynthesis and stomatal conductance of soybean grown at future CO_2 and ozone concentrations for 3 years under fully open-air field conditions. Plant, Cell & Environment, 29, 2077–2090.

Bindi M., Fibbi L., Miglietta F. 2001. Free air CO_2 enrichment (FACE) of grapevine (Vitis vinifera L.): II. Growth and grapes and wine quality in response to elevated CO_2 concentration. Eur .I. Agron. 14, 145–155.

Bindi M., Fibbi L., Frabotta A., Chiesi M., Selvaggi G., Magliulo V. 1999. Free air CO_2 enrichment of potato (Solanum tuberosum L.). In "Annual Report for Changing Climate and Potential Impacts on Potato Yield and Quality (CHIP), Contract ENV4-CT97-0489," pp. 160196. Commission of the European Union, Brussels, Belgium.

Bindi M., Fibbi L., Frabotta A., Ottaviani G., Magliulo V. 1998. Free air CO_2 enrichment of potato (Solanum tuberosum L.). In "Annual Report for Changing Climate and Potential Impacts on Potato Yield and Quality (CHIP), Contract ENV4-CT97-0489," pp. 133–163. Commission of the European Union, Brussels, Belgium.

Bindi M., Fibbi L., Gozzini B., Orlandini S., Miglietta F. 1995. Mini free-air carbon dioxide enrichment (FACE) experiment on grapevine. In "Climate Change and Agriculture in Europe: Assessment of Impacts and Adaptations, Research Report No.9, Contract ENV4-CT97-0489" (P. A. Harrison, R. E. Butterfield, and T. E. Downing, Eds.), pp. 125–137. Commission of the European Union, Brussels, Belgium.

Brooks T. J., Wall G. W., Pinter P. J., Jr., Kimball B. A., LaMorte R. L., Leavitt S. W., Matthias A. D., Adamsen F. J., Hunsaker D. J., Webber A. N. 2001. Acclimation response of spring wheat in a free-air CO_2 enrichment (FACE) atmosphere with variable soil nitrogen regimes. 3. Canopy architecture and gas exchange. Photosyn. Res. 66, 97–108.

Cel W., Czechowska-Kosacka A., Zhang T. 2016. Sustainable mitigation of greenhouse gases emission, Problemy Ekorozwoju – Problems of Sustainable Development, 11(1),173–176.

Cernusak L.A Winter K., Dalling J.W., Holtum J.A.M., Jaramillo C., Körner C., et al. 2013. Tropical forest responses to increasing atmospheric CO_2: currentknowledge and opportunities for future research. Functional Plant Biology, 40, 531–551.

Daepp M., Nosberger J., Liischer A. 2001. Nitrogen fertilization and developmental stage alter ° the response of Lolium perenne to elevated CO_2. New Phytologist 150, 347–358.

Daepp M., Suter D., Almeida J.P.F., Isopp H., Hartwig U., Frehner M., Blum H., Nosberger J., Liischer A. 2000. Yield response of Lolium perenne swards to free-air CO_2 enrichment increased over six years in a high N input system on fertile soil. Global Change Biol. 6, 805–816.

ESRL, Earth System Research Laboratory. https://www.esrl.noaa.gov/gmd/ccgg/trends/data.html, Available online: October 2019.

Garcia R.L., Long S.P., Wall G.W., Osborne C.P., Kimball B.A., Nie G.Y., Pinter P.J., Jr., LaMorte R.L., Wechsung F. 1998. Photosynthesis and conductance of spring wheat leaves: field response to continuous free-air CO_2 enrichment. Plant Cell Environ. 21, 659–669.

GFW, 2018. Global Forest Watch, University of Maryland. www.globalforestwatch.org. Available online October 2019.

Hatch MD. 1999. C4 photosynthesis: a historical overview. In: Sage RF, Monson RK, eds. C4 Plant Biology. San Diego: Academic Press, 17–46.

He H., Kirkham M.B. Lawlor D.J. Kanemasu E.T. 1992. Photosynthesis and water relations of big bluestem (C_4) and kentucky bluegrass (C3) under high concentration carbon dioxide, Transactions of the Kansas Academy of Science, 95(1/2): pp. 139–152.
Hebeisen T., Liischer A., Zanetti S., Fischer B.U., Hartwig U.A., Frehner M., Hendrey G.R., Blum H., Nosberger J. 1997. Growth response of Trifolium repens L. and Lolium perenne L. as monocultures and bi-species mixture to free-air CO_2 enrichment and management. Global Change Biol. 3, 149–160.
Hileman D.R., Huluka G., Kenjige P. K., Sinha N., Bhattacharya N.C., Biswas P.K., Lewin K.F., Nagy J., Hendrey G.R. 1994. Canopy photosynthesis and transpiration of field-grown cotton exposed to free-air CO_2 enrichment (FACE) and differential irrigatiosn. Agric. For. Meteorol. 70, 189–207.
IPCC 2007. The Fourth Assessment Report of the Intergovernmental Panel on Climate Change - The Physical Science Basis. the Carbon Cycle and the Climate System. Contribution of Working Groups I. II and III to the Fourth Assessment Report of the Intergovernmental Panel on Climate Change [Pachauri. R.K. and Reisinger. A. (Eds.). IPCC. Geneva. Switzerland] (https://www.ipcc.ch/publications_and_data/ar4/syr/en/contents.html)
IPCC, 2001. The Third Assessment Report of the Intergovernmental Panel on Climate Change 2001 - The Scientific Basis. Contribution of Working Group I to the Third Assessment Report of the Intergovernmental Panel on Climate Change [Houghton, J.T., Y. Ding, D.J. Griggs, M. Noguer, P.J. van der Linden, X. Dai, K. Maskell, and C.A. Johnson (eds.)]. Cambridge University Press, Cambridge, United Kingdom and New York, NY, USA.
Jongen M., Jones B. 1998. Effects of elevated carbon dioxide on plant biomass production and competition in a simulated neutral grassland community. Annals of Botany, 81(1),111–123. Doi: 10.1006/anbo.1998.0654.
Kim H. Y., Lieffering M., Miura S., Kobayashi K., Okada M. 2001. Growth and nitrogen uptake of CO_2-enriched rice under field conditions. New Phytologist, 150, 223–230.
Kimball B. A., Kobayashi K., Bindi M. 2002. Responses of agricultural crops to free air CO_2 enrichment. Advances in Agronomy, 77, 293–368.
Lara M.V., Andreo C.S. 2011. C4 Plants Adaptation to High Levels of CO2 and to Drought Environments. DOI: 10.5772/24936.
Le Quéré C. Andrew R.A., Friedlingstein P., Sitch S., Pongratz J., Manning A.C., et al. 2018. Global Carbon Budget 2017. Earth Syst. Sci. Data. 10: 405–448.
Leakey A.D.B. 2009. Rising atmospheric carbon dioxide concentration and the future of C4 crops for food and fuel. Proceedings of the Royal Society B, 276, 2333–2343. Doi.org/10.1098/rspb.2008.1517.
Leakey A.D.B., Ainsworth E.A., Bernacchi C.J., Rogers A., Long S.P., Ort P.R. 2009. Elevated CO_2 effects on plant carbon, nitrogen, and water relations: six important lessons from FACE. Journal of Experimental Botany, 60 (10), 2859–2876. Doi:10.1093/jxb/erp096.
Leakey A.D.B., Bishop K.A., Ainsworth E.A. 2012. A multi-biome gap in understanding of crop and ecosystem responses to elevated CO_2. Current Opinion in Plant Biology, 15, 228–236. Doi 10.1016/j.pbi.2012.01.009.
Long S.P. Zhu X.G. Naidu S.L. Ort D.R. 2006. Can improvement in photosynthesis increase crop yields? Plant Cell and Environment 29 315–330.
Ludwig M., 2016. The Roles of Organic Acids in C4 Photosynthesis. Front Plant Sci. 7, 647.
Luo Y., Gerten D., Le Maire G., Parton W.J., Weng E., Zhou X., Keough C., Beier C., Ciais P., Cramer W. et al. 2008. Modeled interactive effects of precipitation, temperature, and [CO_2] on ecosystem carbon and water dynamics in different climatic zones. Global Change Biology, 14, 1986–1999.
Mauney J. R., Kimball B. A., Pinter P. J., Jr., LaMorte R. L., Lewin K. F., Nagy J., Hendrey G. R. 1994. Growth and yield of cotton in response to a free-air carbon dioxide enrichment (FACE) environment. Agric. For. Meteorol. 70, 49–68.
Mauney J. R., Lewin K. F., Hendrey G. R. Kimball B. A. 1992. Growth and yield of cotton exposed to free-air CO_2 enrichment (FACE). Crit. Rev. Plant Sci. 11, 213–222.
Mortensen L.M. 1995. Effect of carbon dioxide concentration on biomass production and partitioning in Betula pubescens Ehrh. seedlings at different ozone and temperature regimes. Environmental Pollution, 87(3),337–343. Doi.org/10.1016/0269-7491(94)P4165-K.
Norby R.J., DeLucia E.H., Gielen B., Calfapietra C., Giardina C.P., King J.S., Ledford J., McCarthy H.R., Moore D.J.P., Ceulemans R. et al. 2005. Forest response to elevated CO_2 is conserved across a broad range of productivity. Proc Natl Acad Sci U S A, 102, 18052–18056.
Osborne C. P., LaRoche J., Garcia R. L., Kimball B. A., Wall G. W., Pinter P. J., Jr., LaMorte R. L., Hendrey G. R., Long S. P. 1998. Does leaf position within a canopy affect acclimation of photosynthesis to elevated CO2? Plant Physiol. 117, 1037–1045.

Ottoman M.J., Kimball B. A., Pinter P.J., Jr., Wall G. W., Vanderlip R.L., Leavitt S.W., LaMorte R.L., Matthias A.D., Brooks T.J. 2001. Elevated CO_2 effects on sorghum growth and yield at high and low soil water content. New Phytologist, 150, 261–273.

Parry M.L. 1990. Climate change and world agriculture. London, Earthscan Publications. http://www.ciesin.org/docs/004-038/004-038a.html.

Portis A.R. Jr Parry M.A.J. 2007 Discoveries in Rubisco (ribulose 1,5-bisphosphate carboxylase/oxygenase): a historical perspective. Photosynthesis Research 94 121–143.

Prior S. A., Rogers H. H., Runion G. B., Mauney J. R. 1994. Effects of free-air CO_2 enrichment on cotton root growth. Agric. For. Meteorol. 70, 69–86.

Prusty B.A.K., Azeez P.A. 2005. Humus: The Natural Organic Matter in the Soil System. J. Agril. Res. & Dev. 1: 1–12.

Rogers A., Fischer B. U., Bryant J., Frehner M., Blum H., Raines C. A., Long, S. P. 1998. Acclimation of photosynthesis to elevated CO2 under low-nitrogen nutrition is affected by the capacity for assimilate utilization. Perennial ryegrass under free-air CO_2 enrichment. Plant Physiol. 118, 683–689.

Rogers H. H., Prior S. A., O'Neill E.G. 1992. Cotton root and rhizosphere responses to free-air CO_2 enrichment. Crit. Rev. Plant Sci. 11, 251–264.

Rosenthal D.M., Tomeo N.J. 2013. Climate, crops and lacking data underlie regional disparities in the CO_2 fertilization effect. Environmental Research Letter, 8, 1–2. Doi:10.1088/1748-9326/8/3/031001.

Sage R.F., Autotrophs, in: Encyclopedia of Ecology, Reference Module in Earth Systems and Environmental Sciences. 2008, s 291–300.

Sneed A. 2018. Ask the experts: Does rising CO_2 benefit plants. https://www.scientificamerican.com/article/ask-the-experts-does-rising-co2-benefit plants1/

Teixeira Da Silva, J.A., Giang, D.D.T. and Tanaka, M. 2006. Photoautotrophic micropropagation of Spathiphyllum. Photosynthetica 44: 53–61.

Tom-Dery D., Eller F., Jensen K., Reisdorff C. 2018. ffects of elevated carbon dioxide and climate change on biomass and nutritive value of Kyasuwa (Cenchrus pedicellatus Trin.). Journal of Applied Botany and Food Quality, 91. https://doi.org/10.5073/JABFQ.2018.091.012.

Van Kessel C., Horwath W. R, Hartwig U., Harris D., and Ltischer A. 2000. Net soil carbon input under ambient and elevated CO2 concentrations: isotopic evidence after four years. Global Change Biol. 6, 435–444.

Wall G. W., Brooks T. J., Adam N. R., Cousins A. B., Kimball B. A., Pinter P. J. Jr., LaMorte R. L., Triggs J., Ottman M. J., Leavitt S. W., Matthias A. D., Williams D. G., Webber, A. N. 2001. Elevated atmospheric CO2 improved Sorghum plant water status by ameliorating the adverse effects of draught. New Phytologist, 152, 231–248.

Wang J., Wang C., Chen N. Xiong Z., Wolfe D. Zou J., 2015. Response of rice production to elevated CO_2 and its interaction with rising temperature or nitrogen supply: a meta-analysis. Climatic Change DOI 10.1007/s10584-015-1374-6.

Wechsung G., Wechsung, F., Wall G. W., Adamsen F. J., Kimball B. A., Pinter P. J., Jr., LaMorte R. L., Garcia R. L., and Kartschall T. 1999. The effects of free-air CO2 enrichment and soil water availability on spatial and seasonal patterns of wheat root growth. Global Change Biol. 5, 519–529.

Zheng Y., Li F., Hao L., Shedayi A.A. Guo L., Ma C. Huang B., Xu M, 2018. The optimal CO_2 concentrations for the growth of three perennial grass species, BMC Plant Biol. 2018; 18: 27.

The Role of Agriculture in Climate Change Mitigation – Pawłowski, Litwińczuk & Zhou (eds)

Methanogenesis and possibilities of reducing it in ruminants

W. Sawicka-Zugaj, W. Chabuz & Z. Litwińczuk
Sub-Department of Cattle Breeding and Genetic Resources Conservation, Institute of Animal Breeding and Biodiversity Conservation, University of Life Sciences in Lublin, Lublin, Poland

1 INTRODUCTION

The human activity has contributed to an increase in the emissions of greenhouse gases, which include carbon dioxide (CO_2), methane (CH_4) and nitrous oxide (N_2O). In consequence, the average annual global temperature in the 20th century increased by 0.4–0.7°C (IPCC 1994, Pathak et al. 2012, Sztumski 2018, Shaw 2010). The main sources of gas emissions are fossil fuel combustion, industry and agriculture. Increased temperature, as one of the elements of climate change, may cause irreversible changes in agriculture, in particular affecting the crop yields, the state of groundwater and soil, as well as livestock farming and fishing (Pathak et al. 2012, Czyżewski et al. 2018).

In animal production, the most greenhouse gases are produced by domestic ruminants, particularly cattle. The main sources of greenhouse gas emissions in ruminant farming are enteric fermentation and storage of natural fertilizers. Enteric fermentation causes the release of methane into the atmosphere, which is considered one of the potential causes of global warming and climate change (Opio et al. 2013, Steinfeld et al. 2006). Ruminants excrete the largest quantity of methane annually (86 Tg of CO_2 equivalent), mainly beef cattle (55.9 Tg), followed by dairy cattle (18.9 Tg) and sheep and goats (9.5 Tg) (McMichael et al. 2007).

According to the National Inventory Report (National Centre for Emissions Management, 2019), the greenhouse gas emissions in Poland totalled 561,772.38 kt of CO_2 eq in 2017, of which 85% came from energy, 6% from industrial processes and product use, and 9% from agriculture. The emission of methane in 2017 alone amounted to 1,976.51 kt (49.41 million tonnes of CO_2 eq), and its share in total domestic greenhouse gas emissions was 11.9%. The three main sources of this gas were fugitive emissions from fuels, agriculture and waste, accounting for 39.7%, 29.3% and 23.2%, respectively. Within agriculture, the largest amount of methane was released by enteric fermentation (25.9%). It is significant that the amount of the methane emitted from agriculture in 2017 had decreased by 35% compared to the baseline year (1988) adopted by the National Centre for Emissions Management, possibly due to a decrease in the number of livestock, especially cattle (from 10,999,400 in 1988 to 6,035,500 in 2017) (National Centre for Emissions Management 2019, Central Statistical Office 2018).

Methanogenesis is a process that leads to the formation of methane, occurring both in the natural environment and as a result of human activity (Table 1).

In agriculture, most CH_4 is emitted via enteric and rumen fermentation (17%), which occurs in the digestive system of ruminants. This is due to the presence and activity of cellulolytic and methanogenic microorganisms. In the fermentation process of structural carbohydrates (cellulose and hemicellulose, contained in the cell wall of plants), these microbes obtain the energy necessary for their metabolism and also bring about the formation of volatile fatty acids (mainly acetate, propionate and butyrate) and the CO_2 and H_2 gases. Methane is released from the gases by methanogenesis: $CO_2 + 8H \rightarrow CH_4 + 2H_2O$ (Brade & Lebzien 2008, Knapp et al. 2014).

1.1 *Mitigation strategies to reduce enteric methane emissions from ruminants*

The work on the identification and quantification of methanogens, as well as on possibilities of limiting methanogens and the methanogenesis process, has been carried out for many years

Table 1. Global CH_4 emissions from natural and anthropogenic sources (Knapp et al. 2014).

Source		% of emission
Natural	Wetlands	30%
	Oceans, lakes & rivers	7%
	Termites & other arthropods	4%
Agricultural	Enteric fermentation	17%
	Rice	6%
	Other agriculture	4%
	Manure	3%
Other anthropogenic	Fossil fuels	15%
	Landfills	6%
	Wastewater	5%
	Biomass burning	3%

(K. A. Johnson & D. E. Johnson 1995, Beauchemin et al. 2009, Buddle et al. 2011, Hill et al. 2016). Because methane is mainly produced as a result of the microbial fermentation of feed components, focus should be placed on controlling the emission of this gas by inhibiting the reaction of hydrogen release or promoting alternative reactions using it (Joblin 1999). A variety of actions can be taken to reduce the methane emission, as presented below. These include the manipulation of rumen fermentation, appropriate genetic selection of animals, changes in animal nutrition, and natural feed additives.

1.2 *Manipulation of rumen fermentation*

It is possible to reduce rumen fermentation and thus the methane production by ruminants by removing protozoa from the rumen, in a process known as defaunation. These microorganisms have a symbiotic relationship with methanogens, taking part in the transfer of hydrogen, which they require to reduce CO_2 to CH_4 (Machmüller et al. 2003). This can be achieved using copper sulphate, acids, surfactants, triazine, lipids, tannins, ionophores or saponins (Hobson & Stewart 1997). In his research, Hegarty (1999) observed that defaunation reduced the methane production by 13%, but the level of reduction depended on the type of diet. The best results were obtained using a high-concentrate diet. Elimination of the protozoa population to mitigate CH_4 release is interesting, but as Haque rightly observes (2018), a lack of protozoa in the rumen may impede the digestion of feed and reduce performance.

Ionophores (polyether antibiotics) such as monensin are the agents that could be used to manipulate rumen fermentation. These are antibacterial agents used commercially in beef and dairy farming to improve productivity (milk and carcass yield), modulate feed consumption, and control flatulence (McGuffey et al. 2001).

The anti-methanogenic effect of monensin involves reducing the number of protozoa in the rumen and increasing the ratio of acetic acid to propionic acid during rumen fermentation by increasing the reducing equivalents that aid in the formation of propionate (Beauchemin et al. 2008). Unfortunately, the growing antibiotic resistance observed in bacteria can also be seen in the case of ionophores, which means that their negative effect on methanogenesis may be short-lived (Johnson & Johnson 1995). In a study by Guan et al. (2006), the addition of monensin to feed (33 mg/kg) caused a 30% reduction in the methane emissions, but only for a period of two months, after which the protozoa adapted to the antibiotics. In addition, the increasing public pressure to reduce the use of bactericides in animal production and the ban on their use in the European Union suggests that the use of monensins is not a long-term means of reducing the methane emissions (Beauchemin et al. 2008, Haque 2018).

Immunization of animals against their own methanogens and protozoa is safe for animals and humans and presents a challenge for the future in working out means to reduce CH_4 production (Clark 2013). The new technique of artificial immunity is aimed at increasing the efficiency of nutrient utilization in farmed ruminants and thereby reducing the methane emissions

(Buddle et al. 2011, Clark et al. 2011). Wright et al. (2004) developed two vaccines for sheep, called VF3 (based on three methanogen strains) and VF7 (seven strains), which reduced the methane production by 7.7% per animal. Cattle were experimentally vaccinated with a methanogenic protein that can produce specific antibodies in both serum and saliva and which is then supplied to the rumen. They remain relatively stable for 4–8 hours and are continually replenished. However, this research has not yet precisely determined what level of antibodies is needed to reduce the methane emissions (Subharat et al. 2015).

1.3 *Genetic selection of animals*

The amount of methane emitted by a herd of animals depends on its size. However, when CH4 production per unit of animal product is considered, the best results are obtained using more productive individuals (Patra 2012, Weisbjerg et al. 2012). The animal performance can be increased through selection towards the amount of product obtained. Kirchgessner et al. (1995) observed that an increase in milk yield from 4000 to 5000 kg/cow/year reduces the methane emissions per kg of raw material by 0.16 for a cow with a body weight of 600 kg. This is also confirmed by later studies (Johnson et al. 1996, Boadi et al. 2004, Beauchemin et al. 2008, Pinares-Patiño et al. 2009; Clark 2013). The Environmental Protection Agency has confirmed that the most promising method of reducing CH_4 from animals is to improve their performance (EPA 2005). An increase in the productivity of dairy cattle may be associated with a decrease in the population of these animals. As demonstrated by Capper et al. (2009), a 400% increase in milk yield in North America resulted in a 64% reduction in the cattle population and a 57% reduction in the methane emissions.

Suppositions that the CH_4 emissions depend on animal size have not been confirmed by research (Tyrrell et al. 1991). A significant factor, however, is the approach to genetic selection for health, disease resistance or reproduction, which definitely affects the life expectancy of animals, their lifetime productivity, and the amount of methane released per unit of product (Buddle et al. 2011).

In addition to increasing productivity, focus should also be placed on the genetic selection of animals for feed conversion efficiency. Hegarty (2001) conducted the research showing that animals' natural variation in the amount of feed consumed per unit of weight gain can be exploited for the breeding purposes by choosing those that consume less feed while still achieving the desired growth rate. At the same time, these animals will produce less methane (Arthur et al. 2001, Clark 2013). Similarly, Yan et al. (2010) and Nkrumah et al. (2006) found that the use of individual dairy cows capable of efficiently utilizing the feed energy for a high level of milk production is a good approach to reducing the CH_4 emissions from lactating dairy cows.

Therefore, it seems that the animal breeding based on genetic selection to obtain low methane emissions is quite feasible, provided it does not negatively affect other production traits.

1.4 *Modifications of animal nutrition*

The simplest solution to reduce the methane emissions by ruminants is to modify their diet in an appropriate manner, which can also increase their productivity. Depending on the extent of their effect and how it is achieved, the diet modifications can achieve a 40% to 75% reduction in emissions (Benchaar et al. 2001, Mosier et al. 1998). Consideration of various feeding strategies should take into account the type of carbohydrate contained in the feed, the level of feed intake, the species of plants used for feed production and their maturity, feeding frequency, processing and preservation of feed, and grazing management (Boadi et al. 2004).

In the case of the type of carbohydrate, fermentation of structural carbohydrates has been clearly shown to cause a greater loss of gross energy intake in the form of CH_4 than the fermentation of easily soluble sugars and starches. This is the effect of a reduced fermentation rate, which in turn results in a higher ratio of acetic acid to propionic acid. In contrast, high levels of consumption of high-grain feeds are associated with high digestion rates and promote greater production of propionic acid. Furthermore, grain-rich feed lowers the rumen pH, which inhibits the growth of methanogens and protozoa (Hegarty & Gerdes 1998, Hegarty 1999). A similar comparison can be made between grass silage and maize silage. As grass for silage is harvested in late maturity, these plants have a lower content of digestible organic matter, sugar and nitrogen than pasture grass and also have a lactate fraction formed in the ensilaging process, which results in higher CH_4 emissions (Tamminga et al. 2007). Maize silage, which is present in the diet of cattle, especially dairy cows, usually has a higher content of both dry matter and easily digestible carbohydrates, which increases digestibility while reducing the methane production. Three facts are important here. First, higher starch content promotes the production of propionate instead of acetate. Secondly, the increased intake of dry matter reduces its time in the rumen, so that fermentation is limited and post-ruminal digestion takes place. Thirdly, the use of this feed increases performance, which translates into lower CH_4 emissions per unit of animal product (O'Mara et al. 1998).

The methane emission also decreases when the feeding level increases. This is mainly due to the rapid passage of feed from the rumen to further sections of the digestive tract and an increased rate of passage through the entire digestive tract. The access of microbes to organic matter is then reduced, which in turn slows down fermentation in the rumen (Mathison et al. 1998, Hegarty 2001). Dairy cows are the most demanding in terms of feeding level, with a requirement for nutrients that far exceeds their physiological capacity for intake. For this reason, properly balanced concentrates are used in the diet of these animals. These contain less fibre, which promotes the production of propionic acid, thereby reducing the CH_4 emissions (Martin et al. 2010). However, it should be noted that in the long term, low fibre content in feed disrupts the rumen function, leading to a serious metabolic disease in the form of subacute or acute acidosis (Owens et al. 1998).

The quality of feed and the frequency of feeding have a decisive impact on the amount of methane produced by cattle. First of all, the importance of the maturity of the plants included in feed should be noted. Studies have clearly shown that the fodder consisting of young plants can reduce the CH4 production due to the lower content of crude fibre and thus the higher proportion of easily digestible carbohydrates, which leads to higher digestibility (Benchaar et al. 2001). Robertson and Waghorn (2002) noted higher CH_4 production from dairy cows grazing in spring compared to those grazing in the summer of the same year. Another factor is the type of plants used as feed. Benchaar et al. (2001), in a model approach to assessing the effectiveness of various existing feeding strategies in reducing the methane production from ruminants, found significantly lower gas production (−21%) in the case of legumes (alfalfa) than for grasses (timothy grass). This effect can be attributed to a lower share of structural carbohydrates in legumes and faster feed retention, which shifts fermentation towards the production of propionate (Johnson & Johnson 1995). In addition, high-sugar grasses such as Lolium multiflorum and Festulolium have been highly successful in reducing methanogens (Buddle et al. 2011).

The amount of methane released during rumen fermentation is also affected by the way the feed is prepared.

The feed that has been ensilaged, known as silage, usually results in lower methanogenesis, because these feeds are already partially fermented (Boadi et al. 2004). In addition, comminution of the feed through cutting or pelleting can reduce CH_4 emissions per kg of dry matter consumed, as smaller particles require less degradation in the rumen (Boadi et al. 2004).

1.5 *Natural feed additives*

Given the high biodiversity of the world of plants and the resulting possibilities for selecting plants for feeding animals, we can focus on the plant species containing chemical compounds that can reduce methanogenesis. These compounds include saponins, natural vegetable tannins and essential oils (Kobayashi 2010).

Saponins are naturally occurring surface-active glycosides produced mainly by plants (Yoshiki et al. 1998). They consist of a sugar moiety, usually composed of glucose, glucuronic acid, xylose, rhamnose or methylpentose (Francis et al. 2002). Research has shown the harmful effects of saponins on protozoa through degradation of cell membranes, and they have been recognized as potential ruminal defaunation agents (Newbold et al. 1997, Wang et al. 2000a). An in vitro study by Guo et al. (2008) has shown that saponins inhibit the growth of protozoa and also limit the capacity of hydrogen for methanogenesis. Many in vivo and in vitro studies (Hess et al. 2003a, Hess et al. 2003b, Hess et al. 2004, Hu et al. 2005, Zeleke et al. 2006) using plants containing saponins, e.g. Acacia angustissima, Camellia sinensis (tea seeds) or the fruits of Sapindus saponaria, Sesbania sesban, Yucca schidigera, and Yucca schidigera, have shown a reduction in methane secretion. However, it should be noted that not all saponins act in the same way (Hess et al. 2003a).

Tannins are polyphenols of varying molecular weight (from several hundred to several thousand) occurring naturally in plants as products of their secondary metabolism. They have the ability to bind to macromolecules, i.e. proteins, structural carbohydrates and starch, thus reducing their digestion (Moumen et al. 2008). According to Tavendale et al. (2005), tannins also reduce the availability of hydrogen, which directly inhibits methanogens and thus reduces methanogenesis. As in the case of saponins, many analytical attempts have been made (Carulla et al. 2005, Grainger et al. 2008, Hess et al. 2003b, Pinares-Pati˜no et al. 2003, Tavendale et al. 2005, Waghorn et al. 2002, Woodward et al. 2004) to confirm this phenomenon in the case of various plants used in ruminant diets (*Acacia mearnsii, Agelaea obliqua, Calliandra calothyrsus, Leucaena leucocephala, Lotus corniculatus, Lotus pedunculatus, Mangifera indica, Medicago sativa, Onobrychis viciifolia,* and *Phylantus discoideus*).

Essential oils, which are volatile aromatic compounds, have very strong antimicrobial properties inhibiting the growth and survival of microbes in the rumen (Greathead 2003, Burt 2004, Tamminga et al. 2004). In this way, essential oils can reduce the amount of hydrogen available for methanogenesis, as confirmed by Grainger et al. (2010), who included cotton seeds in the diet of dairy cows, and by McGinn et al. (2004), who supplemented the diet of beef cattle with sunflower oil. In those studies, the methane emissions were reduced by 17.1% and 21.5%, respectively. In an in vitro study, Chuntrakort et al. (2011) showed that the addition of cotton seed, sunflower seeds and coconut kernels reduced the methane emissions by about 9.0% compared to the control diet. However, it should be noted that although essential oils have proven effective in reducing methanogenesis in the rumen, in some cases they have also adversely affected fibre digestion and fermentation (Macheboeuf et al. 2008, Capper & Buaman 2013), which suggests that further research is necessary.

Attempts have also been made to limit methanogenesis in enteric fermentation in ruminants by supplementing their diet with fats or organic acids. The addition of fats is safer for the animal because, unlike concentrates, they reduce methanogenesis in the rumen without lowering its pH (Sejian et al. 2011). In the experiments carried out by Dohme et al. (2000), the greatest reduction in CH_4 production was obtained using medium-chain fatty acids (C8 – C16) contained in palm kernel oil, coconut oil or high-laurate rapeseed oil, which supplied at the appropriate level reduced the methane emissions by 34%, 21% or 20%, respectively. From a practical point of view, however, Beauchemin et al. (2008) note that the use of medium-chain fatty acids contained in refined oils may not be feasible due to high costs.

A cheaper alternative seems to be long-chain fatty acids contained in oilseeds and animal fat, which do not undergo fermentation. Van der Honing et al. (1981) observed that the addition of 5% animal tallow or soybean oil to the diet of dairy cows reduced methane emissions by 10% and 15%, respectively. When whole oilseeds are used, they should be mechanically treated before being fed to animals (Beauchemin et al. 2008). In the case of supplementation with fats, it is important to use the correct amount, because in excess they slow down the breakdown of fibre in the rumen and reduce the acetate production and the fat content in milk (Mathison et al. 1998, Ashes et al. 1997).

Organic acids have been shown to function as an alternative H_2 sink, and thus factors inhibiting methanogenesis (Castillo et al. 2004). Research (Newbold et al. 2005, McAllister & Newbold 2008) shows that the best of these compounds is fumaric acid, present in mosses and fungi, which is an intermediate compound in the propionic acid pathway, where it is reduced to succinic acid (Boadi et al. 2004). An experiment by Bayaru et al. (2001) in which fumaric acid was added to sorghum silage for Holstein cattle achieved a 23% reduction in the methane production by increasing the production of propionic acid, without affecting the digestibility of dry matter. An additional benefit of using acid in dairy cows is enhanced synthesis of milk protein, resulting from the increased production of propionic acid (Itabashi 2001).

2 CONCLUSIONS

Livestock production, as one of the main sources of anthropogenic greenhouse gases, is a serious problem in many countries around the world. Time and financial resources are increasingly devoted to reducing these emissions. The possibilities under consideration vary in terms of the stage of research, costs, and potential practical applications. The techniques for manipulating methanogens in the rumen cover a very wide range: genetic modifications, manipulation of rumen fermentation, diet modification, or the use of chemical inhibitors. The choice of means to reduce CH_4 in animal production should take into account public sensitivity to the presence of trace amounts of chemicals originating in animal feed in products, especially milk and meat. Hence, the use of chemical inhibitors in the form of ionophores will not gain wide acceptance until this method has been refined. The use of vaccines or genetic selection seems promising; however, we must be aware not only of the costly and time-consuming nature of such research, but also of the impact of numerous environmental factors on the final result.

The health status and production level of animals are important as well. The huge improvement in animal performance over the last century not only presents an opportunity to meet the global milk demand, which is projected to increase by 58% by 2050, but also can be a means of reducing enteric emissions of CH_4 and other greenhouse gases per unit of product (FAO 2011, Knapp et al. 2014).

The processes aimed at defaunation of the rumen environment or at shifting fermentation towards the production of propionic acid may pose a threat to the functioning of the digestive system of ruminants. It should be kept in mind that the microorganisms present in the rumen play two basic roles: they break down feed and are also a source of animal protein, which herbivores cannot consume directly.

At present, the safest solution for humans and animals as well as for a sustainable environment consists in properly refined feeding strategies, including the appropriate species composition of feed, chemical composition (type of carbohydrates contained in the feed), feed quality, feed digestibility, method of feed processing and preservation, and grazing management.

Because these techniques are the simplest to implement, they should become a priority in the research on reducing methane emissions, as well as the main element of knowledge disseminated among livestock producers. This pertains in particular to the use of natural dietary additives, such as herbs containing tannins, saponins or essential oils; fats and organic acids; and high-energy (high-sugar) grasses.

REFERENCES

Arthur, P.F., Archer, J.A., Johnston, D.J., Herd, R.M., Richardson, E.C., Parnell, P. 2001. Genetic and phenotypic variance and covariance components for feed intake, feed efficiency and other postweaning traits in Angus cattle. *Journal of Animal Science* 79: 2805–2811.

Ashes, J. R., Gulati, S. K., Scott, T. W. 1997. New approaches to changing milk composition: Potential to alter the content and composition of milk fat through nutrition. *The Journal of Dairy Science* 80: 2204–2212.

Bayaru, E., Kanda, S., Toshihiko, K., Hisao, I., Andoh, S., Nishida, T., Ishida, M., Itoh, T., Nagara, K., Isobe, Y. 2001. Effect of fumaric acid on methane production, rumen fermentation and digestibility of cattle fed roughages alone. *Journal of Animal Science* 72: 139–146.

Buddle, B.M., Denis, M., Attwood, G.T., Altermann, E., Janssen, P.H., Ronimus, R.S., Pinares-Patiño C.S., Muetzel, S., Wedlock, D.N. 2011. Strategies to reduce methane emissions from farmed ruminants grazing on pasture. *The Veterinary Journal* 188: 11–17.

Beauchemin, K. A., Kreuzer, M., O'Mara, F., McAllister, T. A. 2008. Nutritional management for enteric methane abatement: a review. *Australian Journal of Experimental Agriculture* 48: 21–27.

Beauchemin, K. A., McAllister, T. A., McGinn, S. M. 2009. Dietary mitigation of enteric methane from cattle. CAB Reviews. Perspectives in Agriculture, Veterinary Science, *Nutrition and Natural Resources* 4 (9): 1–18.

Benchaar, C., Pomar, C., Chiquette, J. 2001. Evaluation of dietary strategies to reduce methane production in ruminants: a modelling approach. *Canadian Journal of Animal Science* 81: 563–74.

Boadi, D., Benchaar, C., Chiquette, J., Massé, D. 2004. Mitigation strategies to reduce enteric methane emissions from dairy cows: Update review. *Canadian Journal of Animal Science* 84(3): 319–335.
Brade, W., Lebzien, P. 2008. Reduzierungspotentiale für treibhausgase in der tierernährung und tierhaltung. WBMELV Moderne Tiernährung - sicher, effizient und klimaschonend, Tagungsband, Braunschweig. Bonn: BMELV. 13/14: 45–46.
Buddle, B. M., Denis, M., Attwood, G. T.; Altermann, E., Janssen, P. H., Ronimus, R. S.; Pinares-Patiño, C. S., Muetzel, S., Wedlock, D.N. 2011. Strategies to reduce methane emissions from farmed ruminants grazing on pasture. *The Veterinary Journal* 188 (1): 11–17.
Burt, S. 2004. Essential oils: their antibacterial properties and potential applications in foods - a review. *International Journal of Food Microbiology* 94: 223–53.
Capper, J.L., Bauman, D.E. 2013. The role of productivity in improving the environmental sustainability of ruminant production systems. *The Annual Review of Animal Biosciences* 1 (9), 1–21.
Capper, J. L., Cady, R. A., Bauman, D. E. 2009. The environmental impact of dairy production: 1944 compared with 2007. *Journal of Animal Science* 87: 2160–2167.
Carulla, J.E., Kreuzer, M., Machm¨uller, A., Hess, H.D. 2005. Supplementation of Acacia mearnsii tannins decreases methanogenensis and urinary nitrogen in forage-fed sheep. *Australian Journal of Agricultural Research* 56: 961–970.
Castillo, C., Benedito, J.L., Mendez, J., Pereira, V., Lopez-Alonso, M., Miranda, M., Hernandez, J. 2004. Organic acids as a substitute for monensin in diets for beef cattle. *Animal Feed Science and Technology* 115: 101–16.
Central Statistical Office. 2018. *Statistical Yearbook of Agriculture*. Warsaw.
Chuntrakort, P., Otsuka, M., Hayashi, K., Sommart, K. 2011. Effects of oil plant use for rumen methane mitigation in in vitro gas production. *Khon Kaen Agriculture Journal Suppl.* 39: 246–250.
Clark, H. 2013. Nutritional and host effects on methanogenesis in the grazing ruminant. *Animal Suppl.* 7, 41–48.
Clark, H., Kelliher, F.M., Pinares-Patiño, C.S. 2011. Reducing CH4 emissions from grazing ruminants in New Zealand: challenges and opportunities. *Journal of Animal Science* 24 (2): 295–302.
Czyżewski, B., Guth, M., Matuszczak, A. 2018. The impact of the CAP Green Programmes on farm productivity and its social contribution, *Problemy Ekorozwoju/Problems of Sustainable Development* 13(1): 173–183.
Dohme, F., Machmuller, A., Wasserfallen, A., Kreuzer, M. 2000. Comparative efficiency of various fats rich in medium chain fatty acids to suppress ruminal methanogenesis as measured with RUSITEC. *Canadian Journal Of Animal Science* 80: 473–482.
EPA (Environmental Protection Agency). 2005. Opportunities to Reduce Anthropogenic Methane Emissions in the United States. Publication 430-R-93-012. EPA, Washington, DC.
FAO (Food and Agriculture Organization of the United Nations). 2011. World Livestock 2011: Livestock in food security. FAO, Rome, Italy.
Francis, G., Kerem, Z., Makkar, H.P.S., Becker, K. 2002. The biological action of saponins in animal systems: a review. *British Journal of Nutrition* 88: 587–605.
Grainger, C., Clarke, T., Beauchemin, K.A., McGinn, S.M., Eckard, R.J. 2008. Supplementation with whole cottonseed reduces methane emissions and increases milk production of dairy cows offered a forage and cereal grain diet. *Australian Journal of Experimental Agriculture* 48: 73–76.
Grainger, C., Williams, R., Clarke, T., Wright, A.D.G., Eckard, R.J. 2010. Supplementation with whole cotton seed causes long-term reduction of methane emissions from lactating dairy cows offered a forage and cereal grain diet. *Journal of Dairy Science* 93 (6): 2612–2619.
Greathead, H. 2003. Plants and plant extracts for improving animal productivity. *Proceedings of The Nutrition Society* 62: 279–90.
Guan, H., Wittenberg, K.M., Ominski, K.H., Krause, D.O. 2006. Efficacy of ionophores in cattle diets for mitigation of enteric methane. *Journal of Animal Science* 84 (7): 1896–1906.
Guo, Y.Q., Liu, J.X., Lu, Y., Zhu, W.Y., Denman, S.E., McSweeney, C.S. 2008. Effect of tea saponin on methanogenesis, microbial community structure and expression of mcrA gene, in cultures of rumen micro-organisms. *Letters in Applied Microbiology* 47 (5): 421–426.
Haque, N. 2018. Dietary manipulation: a sustainable way to mitigate methane emissions from ruminants. *Journal of Animal Science and Technology* 60:15.
Hegarty, R. S. 1999. Reducing rumen methane emissions through elimination of rumen protozoa. *Australian Journal of Agricultural Research* 50: 1321–1327.
Hegarty, R. S. 2001. Strategies for mitigating methane emissions from livestock- Australian options and opportunities. Pages 31–34 in Proc. *1st International Conference on Greenhouse Gases and Animal Agriculture*. Obihiro, Hokkaido, Japan.

Hegarty, R.S., Gerdes, R. 1998. Hydrogen production and transfer in the rumen. *Recent Advances in Animal Nutrition* 12: 37–44.
Hess, H.D., Kreuzer, M., D'ıaz, T.E., Lascano, C.E., Carulla, J.E., Soliva, C.R., Machmüller, A. 2003a. Saponin rich tropical fruits affect fermentation and methanogenesis in faunated and defaunated rumen fluid. *Animal Feed Science and Technology* 109: 79–94.
Hess, H.D., Monsalve, L.M., Lascano, C.E., Carulla, J.E., Diaz, T.E., Kreuzer, M. 2003b. Supplementation of a tropical grass diet with forage legumes and Sapindus saponaria fruits: effects on in vitro ruminal nitrogen turnover and methanogenesis. *Australian Journal of Agricultural Research* 54: 703–713.
Hess, H.D., Beuret, R., L¨otscher, M., Hindrichsen, I.K., Machm¨uller, A., Carulla, J. E., Lascano, C. E., Kreuzer, M. 2004. Ruminal fermentation, methanogenesis and nitrogen utilization of sheep receiving tropical grass hay-concentrate diets offered with Sapindus saponaria fruits and Cratylia argentea foliage. *Animal Science* 79: 177–189.
Hill, J., McSweeney, C., Wright, A. G., Bishop-Hurley, G., Kalantar-Zadeh, K. 2016. Measuring Methane Production from Ruminants. *Trends in Biotechnology* 34 (1): 26–35.
Hobson, P. N., Stewart, C. S. 1997. The Rumen Microbial Ecosystem. London: Chapman and Hall.
Hook, S. E., Wright, A.D. G., McBride, B.W. 2010. Methanogens: Methane Producers of the Rumen and Mitigation Strategies. Archaea, Article ID 945785, 11 pages.
Hu, W., Liu, J., Ye, J., Wu, Y., Guo, Y. 2005. Effect of tea saponin on rumen fermentation in vitro. *Animal Feed Science and Technology* 120: 333–339.
Intergovernmental Panel on Climate Change 1994. Radiative Forcing of Climate Change. The 1994 Report of Scientific Assessment. Working Group of IPCC WMO. UNEP. 1–28.
Itabashi, H. 2001. Reducing ruminal methane production by chemical and biological manipulation. Pages 106–111 in Proc. 1st International Conference on Greenhouse Gases and Animal Agriculture. Obihiro, Hokkaido, Japan.
Joblin, K.N. 1999. Ruminal acetogens and their potential to lower ruminant methane emissions. *Australian Journal of Agricultural Research* 50: 1307–1313.
Johnson, K. A., Johnson, D. E. 1995. Methane emissions from cattle. *Journal of Animal Science.* 73 (8): 2483–2492.
Johnson, D. E., Ward, G. M., Ramsey, J. J. 1996. Livestock methane: Current emissions and mitigation potential. In Ervin T. Kornegay (ed.) Nutrient Management of Food Animals to Enhance and Protect the Environment Lewis Publishers CRC Press Inc., Boca Raton, FL: 219–233.
Kirchgessner, M., Windisch, W., Mu¨ller, H.L. 1995. Nutritional factors for the quantification of methane production. In: von Engelhardt, Leonherd-Marke S., Breves G., Giesecke D. (ed.) Ruminant physiology: digestion, metabolism, growth and production. Proceedings of the 8th international symposium on ruminant physiology. Delmar Publishers, Albany, Germany, 333–334.
Knapp, J.R., Laur, G.L., Vadas, P.A., Weiss, W.P., Tricarico, J.M. 2014. Enteric methane in dairy cattle production: Quantifying the opportunities and impact of reducing emissions. *Journal of Dairy Science* 97: 3231–3261.
Kobayashi, Y. 2010. Abatement of methane production from ruminants: trends in the manipulation of rumen fermentation. *Asian-Australian Journal of Animal Sciences* 23(3): 410–416.
Macheboeuf, D., Morgavi, D.P., Papon, Y., Mousset, J.L., Arturo-Schaan, M. 2008. Dose-response effects of essential oils on in vitro fermentation activity of the rumenmicrobial population. *Animal Feed Science and Technology* 145: 335–350.
Machm¨uller, A., Soliva, C. R., Kreuzer, M. 2003. Effect of coconut oil and defaunation treatment on methanogenesis in sheep. *Reproduction Nutrition Development* 43 (1): 41–55.
Mathison, G. W., Okine, E. K., McAllister, T. A., Dong, Y., Galbraith, J., Dmytruk, O. I. N. 1998. Reducing methane emissions from ruminant animals. *Journal of Applied Animal Research* 14: 1–28.
Martin, C., Morgavi, D.P., Doreau, M. 2010. Methane mitigation in ruminants: from microbe to the farm scale. *Animal* 4: 351–65.
McAllister T.A., Newbold C.J. 2008. Redirecting rumen fermentation to reduce methanogenesis. *Aust J Exp Agric.* 48, 7–13.
McGinn, S.M., Beauchemin, K.A., Coates, T., Colombatto, D. 2004. Methane emissions from beef cattle: effects of monensin, sunflower oil, enzymes, yeast, and fumaric acid. *Journal of Animal Science* 82 (11): 3346–3356.
McGuffey, R.K., Richardson, L.F., Wilkinson, J.I.D. 2001. Ionophores for dairy cattle: current status and future outlook. *Journal of Dairy Science* 84 (E. Suppl.): 194–203.
McMichael, A. J., Powles, J. W., Butler, C. D., Uauy, R. 2007. Food, livestock production, energy, climate change, and health. *The Lancet* 370 (9594): 1253–1263.

Mosier, A.R., Duxbury, J.M., Freney, J.R., Heinemeyer, O., Minami, K., Johnson, D.E. 1998. Mitigating agricultural emissions of methane. *Climatic Change* 40: 39–80.

Moumen, A., Yáñez-Ruiz, D.R., Martín-García, I., Molina-Alcaide, E. 2008. Fermentation characteristics and microbial growth promoted by diets including two-phase olive cake in continuous fermenters. *Journal of Animal Physiology and Animal Nutrition* 92 (1): 9–17.

National Centre for Emissions Management. 2019. Poland's national inventory report 2019. Greenhouse gas inventory for 1988–2017. Warsaw.

Newbold, C.J., Lopez, S., Nelson, N., Ouda, J.O., Wallace, R.J., Moss, A.R. 2005. Propionate precursors and other metabolic intermediates as possible alternative electron acceptors to methanogenesis in ruminal fermentation in vitro. *British Journal of Nutrition* 94: 27–35.

Nkrumah, J.D., Okine, E.K., Mathison, G.W., Schmid, K., Li C., Basarab, J.A., Price, M.A., Wang, Z., Moore, S.S. 2006. Relationships of feedlot efficiency, performance, and feeding behavior with metabolic rate, methane production, and energy partitioning in beef cattle. *Journal of Animal Science* 84: 145–153.

O'Mara, F.P., Fitzgerald, J.J., Murphy, J.J., Rath, M. 1998. The effect on milk production of replacing grass silage with maize silage in the diet of dairy cows. *Livestock Production Science* 55: 79–87.

Opio, C., Gerber, P., Mottet, A., Falcucci, A., Tempio, G., MacLeod, M., Vellinga, T., Henderson, B., Steinfeld, H. 2013. Greenhouse gas emissions from ruminant supply chains – a global life cycle assessment. Rome, Italy: Food and Agriculture Organization of the United Nations (FAO).

Owens, F.N., Secrist, D. S., Hill, W. J., Gill, D. R. 1998. Acidosis in cattle: a review. *Journal of Animal Science* 76 (1): 275–286.

Pathak, H., Aggarwal, P.K., Singh, S.D. (ed.). 2012. Climate Change Impact, Adaptation and Mitigation in Agriculture: Methodology for Assessment and Applications. *New Delhi: Indian Agricultural Research Institute.*

Patra, A.K. 2012. Enteric methane mitigation technologies for ruminant livestock: a synthesis of current research and future directions. *Environmental Monitoring and Assessment* 184: 1929–1952.

Pinares-Pati˜no, C.S., Ulyatt, M.J., Waghorn, G.C., Lassey, K.R., Barry, T.N., Holmes, C. W., Johnson, D. E. 2003. Methane emission by alpaca and sheep fed on Lucerne hay or grazed on pastures of perennial ryegrass/white clover or birdsfoot trefoil. *Journal of Agricultural Science* 140: 215–226.

Pinares-Patiño, C. S., Waghorn, G. C., Hegarty, R. S., Hoskin, S. O. 2009. Effects of intensification of pastoral farming on greenhouse gas emissions in New Zealand. *New Zealand Veterinary Journal* 57: 252–261.

Robertson, L. J., Waghorn, G. C. 2002. Dairy industry perspectives on methane emissions and production from cattle fed pasture or total mixed rations in New Zealand. *Proceedings of the New Zealand Society of Animal Production* 62: 213–218.

Sejian, V., Lakritz, J., Ezeji, T., Lal, R. 2011. Forage and flax seed impact on enteric methane emission in dairy cows. *Research Journal of Veterinary Sciences* 4 (1): 1–8.

Shaw, K. 2019. Implementing Sustainability in Global Supply Chain, *Problemy Ekorozwoju/Problems of Sustainable Development* 14(2):117: 127.

Steinfeld, H., Gerber, P., Wassenaar, T., Castel, V., Rosales, M., Haan, C.D. 2006. Livestock's long shadow: Environmental issues and options. Rome, *Italy: Food and Agriculture Organization of the United Nations* (FAO).

Subharat, S., Shu, D., Zheng, T., Buddle, B. M., Janssen, P.H., Luo, D., Wedlock, D.N. 2015. Vaccination of cattle with a methanogen protein produces specific antibodies in the saliva which are stable in the rumen. *Veterinary Immunology and Immuno*pathology 164: 201–207.

Sztumski, W. 2018. Responsible development and durable development, *Problemy Ekorozwoju/Problems of Sustainable Development* 13(1):113: 120.

Tavendale, M.H., Meagher, L.P., Pacheco, D., Walker, N., Attwood, G.T., Sivakumaran, S. 2005. Methane production from in vitro rumen incubations with Lotus pedunculatus and Medicago sativa, and effects of extractable condensed tannin fractions on methanogenesis. *Animal Feed Science and Technology* 123–124 (1): 403–419.

Tamminga, S., Bannink, A., Dijkstra, J., Zom, R. 2007. Feeding strategies to reduce methane loss in cattle. Lelystad: *The Netherlands: Animal Nutrition and Animal Sciences Group*, Wageningen UR, Report.

Tyrrell, H. F., Reynolds, C. K., Blaxter, H. D. 1991. Utilization of dietary energy by Jersey compared to Holstein cows during the lactation cycle. Proc. 12th Symp. Energy Metabolism of Farm Animals. EAAP Publ. 58. *Rome: European Federation for Animal Science (EAAP).*

Van der Honing, Y., Weiman, B. J., Steg, A., Van Donselaar, B. 1981. The effect of fat supplementation of concentrates on digestion and utilization of energy by productive dairy cattle. *Netherlands Journal of Agricultural Science* 29: 79–85.

Waghorn, G.C., Tavendale, M.H., Woodfield, D.R. 2002. Methanogenesis from forages fed to sheep. *Proceedings of the New Zealand Grassland Association* 64: 159–165.
Wang, Y., McAllister, T.A., Yanke, L.J. Cheeke, P.R. 2000. Effect of steroidal saponin from Yucca schidigera extract on ruminal microbes. *Journal of Applied Microbiology* 88: 887–896.
Weisbjerg, M.R., Terkelsen, M., Hvelplund, T., Madsen, J. 2012. Increased productivity in tanzanian cattle production is the main approach to reduce methane emission per unit of product. in Book of Abstracts, *35th Annual scientific conference*. Olasiti Garden, Arusha, Tanzania.
Woodward, S.L., Waghorn, G.C., Laboyre, P. 2004. Condensed tannins in birdsfoot trefoil (Lotus corniculatus) reduced methane emissions from dairy cow. *Proceedings of the New Zealand Society of Animal Production* 64: 160–164.
Wright, A.D.G., Kennedy, P., O'Neill, C.J., Toovey, A.F., Popovski, S., Rea, S.M., Pimm, C.L., Klein, L. 2004. Reducing methane emissions in sheep by immunization against rumen methanogens. *Vaccine* 22 (29–30): 3976–3985.
Yan, T., Mayne, C.S., Gordon, F.G., Porter, M.G., Agnew, R.E., Patterson, D.C., Ferris, C.P., Kilpatrick, D.J. 2010. Mitigation of enteric methane emissions through improving efficiency of energy utilization and productivity in lactating dairy cows. *Journal of Dairy Science* 93 (6): 2630–2638.
Yoshiki, Y., Kudou, S., Okubo, K,. 1998. Relationship between chemical structures and biological activities of triterpenoid saponins from soybean (Review). *Bioscience Biotechnology and Biochemistry* 62: 2291–2299.
Zeleke, A.B., Cl′ement, C., Hess, H.D., Kreuzer, M., Soliva, C.R. 2006. Effect of foliage from multipurpose trees and a leguminous crop residue on in vitro methanogenesis and ruminal N use. In Carla R. Soliva, J. Takahashi, M. Kreuzer (ed.) *Greenhouse gases and animal agriculture: an update*. Elsevier International Congress Series 1293: 168–171.

The Role of Agriculture in Climate Change Mitigation – Pawłowski, Litwińczuk & Zhou (eds)
© 2020 Taylor & Francis Group, London, ISBN 978-0-367-43372-7

Improving furfural tolerance of recombinant E. coli in the fermentation of lignocellulosic sugars into ethanol

Z. Huabao, X. Shuangyan & Z. Tao
Key Laboratory of Soil Contamination Bioremediation of Zhejiang Province, College of Environmental and Resource Sciences, Zhejiang A&F University, Hangzhou, China

1 INTRODUCTION

Lignocellulose is an abundant renewable resource that can be converted into fuels and chemicals by biocatalysts. In contrast to starch, sugarcane, and sugar beet, the use of lignocellulosic residues and short rotation trees would not directly compete with food production (Liu et al. 2019, Żukowska et al. 2016). Unlike starch, lignocellulose has been designed by nature to resist deconstruction. Crystalline fibers of cellulose are encased in a covalently linked mesh of lignin and hemicellulose. Dilute acid pretreatment of lignocellulose has been widely investigated to increase enzyme access to cellulose and hydrolyze hemicelluloses (Liu & Bao 2019, Pimentel 2012, Dowbor 2013). During this acid pretreatment, small amounts of side products including furans, carboxylic acids, and aromatic compounds are produced that retard microbial fermentation.

Expression of fucO from plasmids has been used to improve furfural tolerance in Escherichia coli-based fermentations for ethanol and lactic acid (Wang et al. 2011). The reduction of furfural to the less toxic alcohol seems essential for growth and fermentation of dilute acid hydrolysates of hemicelluloses (Wang et al. 2019). In this study, we have used site-specific mutagenesis and growth-based selection to identify a fucO mutation that confers a further increase in furfural tolerance.

2 RESEARCH MATERIALS AND METHODS

The strains, plasmids, and primers used in this study are described as listed in Table 1. LB medium containing xylose was used for the construction of ethanol strains. AM1 minimal salts medium with xylose was used for the maintenance and growth of ethanologenic strains. Solid medium contained 20 g/L xylose. Broth cultures contained 50 g/L xylose. Batch fermentations contained 100 g/L xylose. Cultures were incubated at 37°C unless stated otherwise. Plates streaked with ethanologenic strains were incubated under argon.

Standard genetic methods were used for the isolation of DNA and plasmids, digestion with restriction enzymes, PCR amplification of DNA, and plasmid constructions. Enzymes were purchased from New England BioLabs (Ipswich, MA) and used as directed by the vendor. Plasmid constructions were confirmed by Sanger sequencing.

3 RESULTS

Furfural has been shown to cause DNA damage in Escherichia coli and to inhibit growth until furfural has been substantially metabolized to the less toxic alcohol. Strategies have been developed to reduce the toxicity of dilute acid hydrolysates. A chromosomal library of Bacillus subtilis YB886 was screened to identify thyA that increased the furfural tolerance of E. coli LY180 (Zheng et al. 2012). The enzyme L-1, 2-propanediol oxidoreductase (encoded by fucO) is an NADH-linked, iron-dependent group III dehydrogenase.

Table 1. Bacterial strains, plasmids, and primers.

Strains, plasmids or primers	Relevant characteristicsa	Reference
Strains		
TOP10F'	F'[lacIq Tn10(tetR)] mcrA Δ(mrr-hsdRMS-mcrBC) φ80lacZΔM15 ΔlacX74 deoR nupG recA1 araD139 Δ(ara-leu)7697 galU galK rpsL(StrR) endA1 λ-	Invitrogen
XW92	LY180(ΔyqhD)	This study
LY180a	ΔfrdBC::(ZmfrgcelYEc), ldhA::(ZmfrgcasABKo), adhE::(ZmfrgestZPpFRT), ΔackA::FRT, rrlE::(pdcadhAadhBFRT),ΔmgsA::FRT	
Plasmids		
pCR2.1-TOPO	Plac, bla, kan	(Invitrogen)
pTrc99A	Ptrc, bla, lacIq,	(Pharmacia)
pLOI5536	fucOL8F in pTrc99a	This study
Primers		
F-L8F-fuco	GCTAACAGAATGATTTTAACGAAACGGCATGGTTTGGT	This study
R-L8F-fuco	ACCAAACCATGCCGTTTCGTTAAAATCATTCTGTTAGC	This study
fuco-RT-F	ACGCCGTGGTTATCAGAAGG	This study
fuco-RT-R	CAGGTAATCCGCGCCGCTAT	This study

This enzyme has a broad substrate range that includes furfural. Expression of fucO from plasmids has been used to improve furfural tolerance in Escherichia coli-based fermentations for ethanol and lactic acid. We have used site-specific mutagenesis and growth-based selection to identify a fucO mutation L7F that shows an increased specific activity by 10-fold than wild-type fucO (Figure 1A). This increase in specific activity was confirmed from SDS-PAGE (12.5% acrylamide) (Figure 1B). One band of soluble mutated sample in the fucO region (38.2 kDa) was observed more dense than that of wild type fucO.

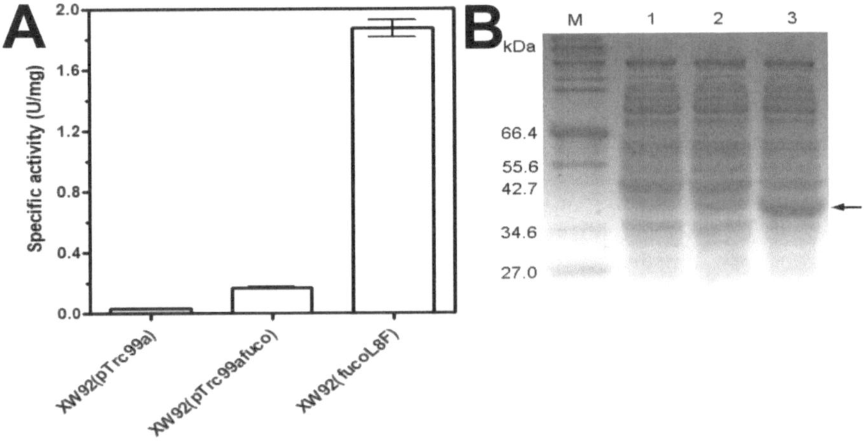

Figure 1. Specific activity of wild type fucO and its mutant fucOL8F (A), pTrc99a is empty vector in which fucO is cloned. SDS-PAGE of crude extract from strains XW92 and pTrc99a (lane 1), pTrc99a derivatives containing fucO genes (lane 2), pTrc99a derivatives containing fucOL8F gene (lane 3) (B). The arrow indicates the putative FucO protein from plasmid expression. M, molecular mass marker lanes.

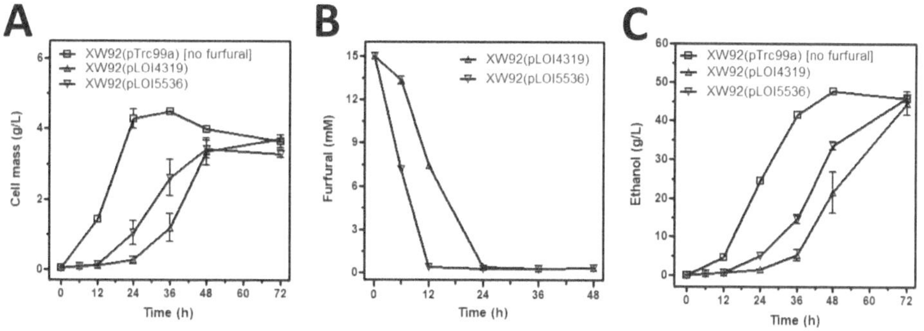

Figure 2. Increased expression of fucO increased the furfural degradation rate in fermentation. (A) Cell mass of fermentation on 15 mM furfural. (B) Furfural degradation of strains during fermentation with 15 mM furfural. (C) Ethanol production of strains during fermentation with 15 mM furfural.

The L7F FucO mutation also improved the fermentation performance (strain XW92) in AM1 medium with 100 g/L xylose and 15mM furfural (Figure 2). With the mutant gene (pLOI5536), 15mM furfural was completely metabolized in 12 h, compared to 24 h with the native fucO gene (pLOI4319) and 48 h with the empty vector. With all two strains, growth and ethanol production were delayed until furfural had been substantially metabolized to the corresponding alcohol. In the presence of 15 mM furfural in broth, both wild-type strain and fucO mutant strain reached the same ethanol titer at 72 h as XW92(pTrc99a) did without furfural.

4 DISCUSSION AND CONCLUSIONS

Many methods are available to create sequence diversity libraries, including chemical mutagenesis, error-prone PCR, saturation mutagenesis, and DNA shuffling. With the mutant gene, furfural was metabolized in vivo at twice the rate of the native enzyme during fermentation (Wang et al. 2011). In this study, we have used site-specific mutagenesis and growth-based selection to identify a fucO mutation that confers a further increase in furfural tolerance. The fucO(L7F) mutant exhibited a 10-fold increase in cytoplasmic activity and was isolated after the screening of only 1,400 colonies. The site of this mutation was unexpected, with the contact region of homodimers distant from the active site. With the mutant gene, furfural was metabolized in vivo at twice the rate of the native enzyme during fermentation.

Future research will be focused on further increasing known enzymes on catalytic activity or exploring new enzymes that were able to degrade furfural. Furthermore, systematic process optimization including strain engineering, biomass pretreatment and fermentation will be carried out to increase the titer of ethanol production in presence of furfural (Liu & Bao 2017, Liu et al. 2018).

ACKNOWLEDGEMENT

This study was supported by the National Key Research and Development Program (2018YFD0500206); and the Research and Development Fund of Zhejiang A&F University (2034020081).

REFERENCES

Dowbor, L., 2013. Economic Democracy - Meeting Some Managmenet Challenges: Changing Scenarios in Brazil. *Problems of Sustainable Development/Problemy Ekorozwoju*, 8(2):17–25.

Liu, C.G., Xiao, Y., Xia, X.X. et al. 2019. Cellulosic ethanol production: Progress, challenges and strategies for solutions. *J. Biotechnol. Adv.* 37(3):491–504.

Liu, G., Bao, J. 2019. Constructing super large scale cellulosic ethanol plant by decentralizing dry acid pretreatment technology into biomass collection depots. *J. Bioresour. Technol.* 275:338–344.

Pimentel, D. 2012. Energy Production from Maize. *Problems of Sustainable Development/Problemy Ekorozwoju*, 7(2):15–22.

Wang, X., Miller, E.N., Yomano, L.P. et al. 2011. Increased furfural tolerance due to overexpression of NADH-dependent oxidoreductase FucO in Escherichia coli strains engineered for the production of ethanol and lactate. *J. Appl. Environ. Microbiol.* 77(15):5132–5140.

Wang, L., York, S.W., Ingram, L.O. et al. 2019. Simultaneous fermentation of biomass-derived sugars to ethanol by a co-culture of an engineered Escherichia coli and Saccharomyces cerevisiae. *J. Bioresour. Technol.* 273:269–276.

Żukowska, G. et al. 2016. Agriculture vs. Alleviating the Climate Change. *Problems of Sustainable Development/Problemy Ekorozwoju* 11(2):67–74.

The Role of Agriculture in Climate Change Mitigation – Pawłowski, Litwińczuk & Zhou (eds)

Simulating carbon sequestration capacity of forests in subtropical area: A case study in Hunan Province, southern China

W. Xiang, M. Zhao, Z. Zhao, P. Lei, S. Ouyang & X. Deng
Huitong National Station for Scientific Observation and Research of Chinese Fir Plantation Ecosystem, Hunan Province, Huitong, China
Faculty of Life Science and Technology, Central South University of Forestry and Technology, Changsha, Hunan, China

X. Zhou & C. Peng
Center of CEF/ESCER, Department of Biological Sciences, University of Quebec at Montreal, Montreal, QC, Canada

1 INTRODUCTION

Forest carbon (C) sequestration involves processes that capture atmospheric carbon dioxide (CO_2) and then store C within forest ecosystems over long periods (FAO 2010; Pan et al. 2011). The capacity for C sequestration by forests is usually assessed by the rate at which CO_2 is removed from the atmosphere (C sink) and/or the quantity of C retained within woodland reservoirs (C storage) (Pan et al. 2011). Between 1990 and 2007, the global forested area was 4.03×103 million ha (FAO 2010), containing 363 Pg C (1 Pg = 1015 g) total stand biomass and absorbing 2.4 ± 0.4 Pg C yr-1, which offset 16% of total fossil-fuel emissions (Pan et al. 2011); however, regional patterns and sensitivities related to C sequestration by forests remain undefined and uncertain.

Accurate estimations of C sequestration in forests on a regional scale have major implications for regional C-balance assessment (Yu et al. 2013), sustainable forest development (Zhou et al. 2013) and multi-scaled data fusion (Fang et al. 2001). Critical tools related to the accurate regional estimation of C sequestration in forests include sufficient data sources, reliable calculation methods and comprehensive factor attributions (Houghton 2005). Field observations and in situ measurements are destructive, costly and spatiotemporally limited (Houghton 2005). Remotely sensed data are lower in cost, allow for large-spatial-scale investigations and employ multi-temporal data and analysis (Yin et al. 2015, Yu et al. 2008), but it is difficult to obtain whole images with accurate spatial resolutions under meteorological conditions using this approach (Piao et al. 2005). Forest inventories are comprehensive, periodic and can attain high estimate accuracy (Chen et al. 2013), but they rarely map spatial patterns (Houghton 2005). The use of inventory, satellite and field data for initialization, parameterization and validation in process-based hybrid models is promising for regional estimation, especially for geographically complex regions (Peng et al. 2009, Zhang et al. 2008, Zhao et al. 2013).

Subtropical forests in southern China cover approximately 25% of the nation's land area (Lin et al. 2012) and play a substantial role in the overall forest C budget, which contributes greatly to Northern Hemisphere terrestrial C-cycling dynamics (Fang et al. 2001, Piao et al. 2011, Zhou et al. 2013). They are also a major potential C sink (Song et al. 2013). The data reported by Yu and Chen et al. (2014) indicated that East-Asian subtropical monsoon forests between 20°N and 40°N have an average net ecosystem productivity (NEP) of 362± 39 g C m^{-2} yr-1 (mean ± 1 S.E.) (Yu et al. 2014). This NEP is higher than those of Asian tropical and temperate forests and also higher than those of forests at the same latitudes in Europe, Africa and North America. However, heterogeneous topography, complex hydrothermal conditions and diverse forest landscapes

(Wang et al. 2013, Zhao et al. 2013) result in not only a scientific challenge but also an opportunity to assess Chinese subtropical C sequestration by forests on a regional scale (Yu et al. 2014, Zhou et al. 2013).

The well-established and tested process-based forest-growth and C dynamic model TRIPLEX1.6 (Peng et al. 2002) was used to estimate C sequestration by forests using key indices of gross primary productivity (GPP), NEP, aboveground and total stand biomass, which spans a broad range of site and climatic conditions through data obtained from the major forest types found throughout Hunan Province. The objectives of this study were: (1) to analyze spatial patterns of forest C storage and sequestration rate; and (2) to carry out sensitivity analysis of C sequestration by forests and evaluate implications for local and regional forest C sequestration capacity dynamics and forest management.

2 MATERIALS AND METHODS

2.1 *Study area*

This study was carried out in Hunan Province (lat 24°38′ to 30°08′ N and long 108°47′ to 114° 15′ E) in central subtropical China. The landform is characterized by mountains in the east, south and west with elevations ranging from 200 to 1800 m, hills in the east-central region and plains in the northeast region surrounding Dongting Lake. The topography is heterogeneous, with slopes varying between 0° and 70°. The total land area is 21.18 million ha, including 12.53 million ha of forest, 3.92 million ha of arable land, 1.2 million ha covered by water bodies and 1.0 million ha of developed land.

This region has a continental and subtropical humid monsoon climate. The annual rainfall is 1550±550 mm, with 75% occurring between March and August. The mean annual temperature is 16.9±1.6°C: 5.3±2.6°C during the coldest month (January) and 28.3±1.2°C during the warmest month (July). Soil types include red soil, yellow soil and red-yellow soil, with small areas of lime soil, purple soil, and paddy soil, which mainly developed from granite, limestone and shale.

Evergreen broadleaved forest is the climax vegetation type in Hunan Province; however, due to anthropogenic disturbances, there are diverse forest types, including plantations and secondary forests. The subtropical forests are generally grouped into four forest types according to the dominant tree species: pine forest, fir forest, deciduous forest and evergreen broadleaved forest. Pine forests accounted for 1.91 million ha in 2004, with tree species including *Pinus massoniana* and *P. elliottii*. Fir forests covered 3.15 million ha in 2004, with tree species including *Cunninghamia lanceolata* and *Cupressus funebris*. Deciduous and evergreen broadleaved forests represent forests at various successional stages, depending upon disturbance intensity and restoration time. The area of deciduous forests was 0.53 million ha in 2004, with tree species including *Liquidambar formosana, Alniphyllum fortune, Choerospondias axillaris*, and *Magnolia denudate*. The area of evergreen broadleaved forests was 1.37 million ha in 2004, with tree species including *Cyclobalanopsis glauca, Lithocarpus glaber, Cinnamomum camphora, Camellia oleifera, Schima superb, Phoebe zhnnan*, and *Ulmus pumila*.

2.2 *TRIPLEX model*

TRIPLEX is a generic hybrid model that can simulate C flux and budgets in forest ecosystems using input data derived from site, soil, climate and stand growth (Peng et al. 2002). The model is composed of four submodels (Peng et al. 2002, Zhou et al. 2004). The forest-production submodel estimates monthly GPP (including above and belowground biomass) from photosynthetic active radiation (PAR), mean air temperature, vapor-pressure deficit (VPD), soil water, the percentage of frost days and the leaf-area index. The forest-growth and yield submodel calculates tree-growth and yield variables (height and tree diameter at breast height) using a function of the stem-wood-biomass increment (Bossel 1996). The soil-C and soil-N submodel simulates soil C and N dynamics between litter and soil pools based upon

CENTURY soil-decomposition modules (Parton et al. 1993). The soil-water-balance submodel, originating from the soil-water submodel of the CENTURY model, simulates water balances and dynamics and calculates monthly water loss through transpiration, evaporation, soil water content and snow water content (Parton et al. 1993).

The TRIPLEX1.6 simulation requires input data such as latitude, longitude, soil texture, monthly climate records, tree-species physiological variables (such as maximum tree height and diameter), tree-species process mediators, stand structures and certain initial site conditions. Simulation outputs include tree diameter, height, basal area, total volume, leaf-area index, GPP, NEP, biomass, soil C, and N and water dynamics, among others. TRIPLEX1.6 has recently been tested and applied to forest growth and biomass production on a regional scale in Zhejiang Province, China (Zhang et al. 2008) and northeastern China (Peng et al. 2009) and was also used to predict *C. lanceolata* and *P. massoniana* stand production in Hunan Province, southern China (Zhao et al. 2013). The studies provided a solid foundation for the application of the TRIPLEX1.6 model in this study.

2.3 *Model input data, calibration, validation and simulations*

There are five primary data sources: permanent forest-plot records (667 m^2) taken from the forest inventory in Hunan Province, the climate dataset, field observations, the literature and assumptions. Field investigations in the permanent forest-plot dataset include records of location, site conditions and diameter at breast height (DBH) of each tree, as well as average stand height, and generates data on land use, site class, dominant tree species, stand density, age, average DBH, average tree height and volume. Stand biomass can be estimated from stand data. Climate data for each permanent plot were interpolated from data collected from meteorological stations (Zhao et al. 2013). The climate inputs were monthly frost days, monthly average air temperature, monthly sums of precipitation and the monthly average atmospheric vapor-pressure deficit (VPD) (Peng et al. 2002). Atmospheric N deposition was set to 18 kg N ha^{-1} $year^{-1}$ (Chen & Mulder 2007). Field observations were made at independent sampling plots in Hunan Province. The choice of reference literature and assumptions were based on indicator functions for the four forest types selected for this study (Xiang et al. 2011).

For site and stand initialization, data were obtained from permanent forest plots in Hunan Province in 2004. Plots were required to meet conditions regarding site, climate, and growth, and the initial data were required to be detailed and representative of the forest types, as well as of the study area as a whole (Zhao et al. 2013). As a result, a total of 1286 plots (427 pine forest plots, 721 fir forest plots, 59 deciduous forest plots and 79 evergreen broadleaved forest plots) were selected in 2004 for initialization of the TRIPLEX1.6 model (Figure 1b). Climate data interpolated from meteorological stations recorded between 2000 and 2009 were used as climate inputs for each plot. To maximize simulation accuracy, specific site and forest-type parameters were determined, calibrated and generalized for several representative plots, locations, sites, regions, forest types or process mediators. Some stand-structure parameters were related to growth variables and exhibited dynamic properties. Consequently, the relationships between crown-diameter and height-diameter were established.

Model validation employed independent literature sources and 2009 permanent forest-plot data. Validation results indicated that model simulations were in good agreement with observations. The TRIPLEX1.6 model was capable of simulating average stand height, DBH, and the above- and belowground biomass of the four forest types (Appendix G, H). It proved to be practical for simulating stand density, average stand DBH and height and above- and belowground biomass on regional scales.

Forest C-sequestration variables (i.e., GPP, NEP, aboveground and total stand biomass) were simulated after TRIPLEX1.6 calibration (Peng et al. 2002) for all selected plots. Plot-based simulations were then scaled up to the provincial scale with a gap-filling method using site, climate and stand structure (Falge et al. 2001).

3 RESULTS

3.1 *Variations in C sequestration in forests*

Average GPP was 7.96±0.21 tC ha^{-1} year^{-1} (ranging from 0.07 to 12.71 tC ha^{-1} yr^{-1}), and NEP was averaged to 1.203±0.05 tC ha^{-1} yr^{-1} with a range from −1.04 to 2.19 tC ha^{-1} yr^{-1}. Average aboveground stand biomass was 27.01±1.28 tC ha^{-1} (ranging between 0.09 and 120.88 tC ha^{-1}), while total stand biomass was 35.87±1.56 tC ha^{-1} (ranging between 0.15 and 141.67 tC ha^{-1}) (Table 1).

Significant and positive correlations were found among spatial patterns of GPP, NEP, aboveground and total biomass C density in forests in Hunan Province ($p<0.01$) (Figure 1).

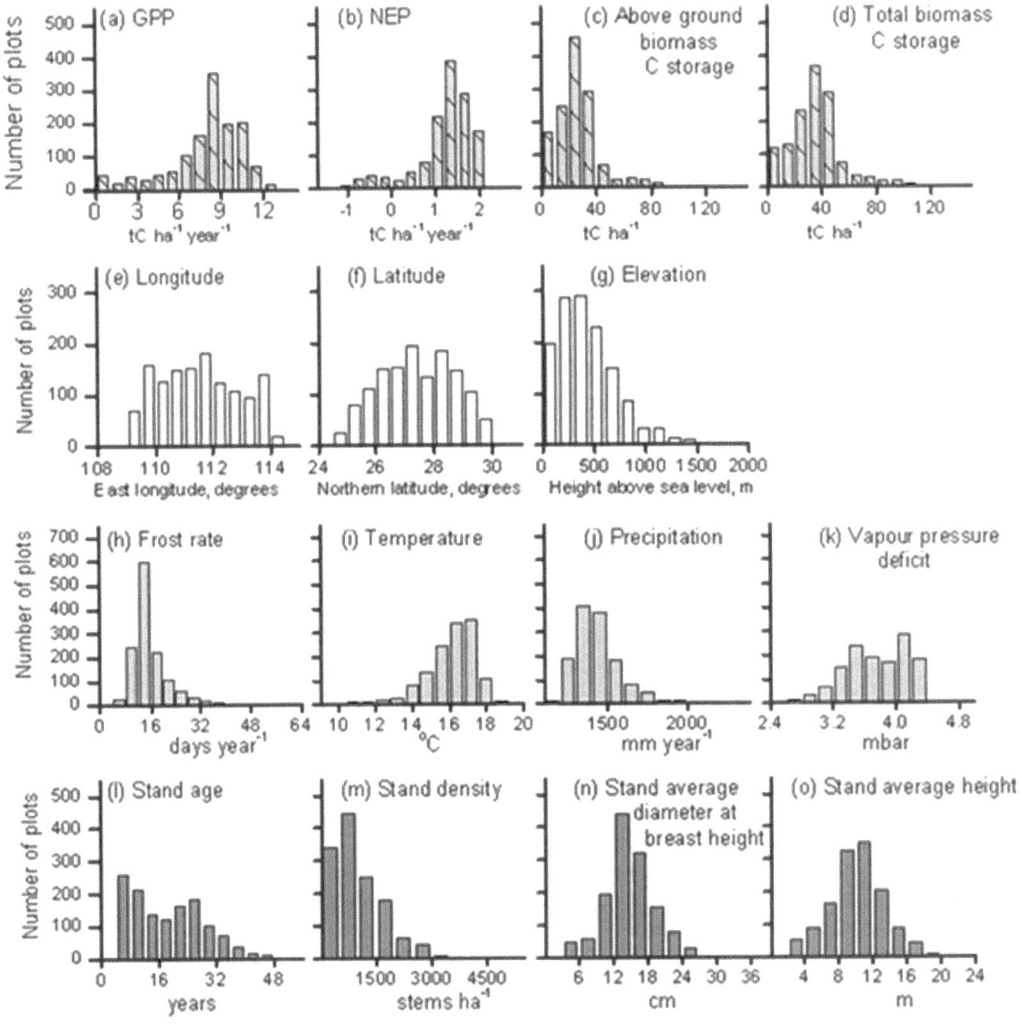

Figure 1. Different class-value distributions of forest C-sequestration variables (GPP (a), NEP (b), aboveground biomass C storage (c) and total biomass C storage (d)) and drivers (longitude (e), latitude (f), elevation (g), annual frost days (h), annual average temperature (i), annual total precipitation (j), the annual average vapor-pressure deficit (k), stand density (l), stand age (m), stand average diameter at breast height (n) and average height (o)) in all selected permanent forest sample plots in Hunan Province in 2009.

Table 1. Descriptive statistics for the spatial heterogeneity of C sequestration by forests in all selected permanent forest plots in Hunan Province in 2009. Forest C sequestration variables include GPP (tC ha^{-1} yr^{-1}), NEP (tC ha^{-1} $year^{-1}$), aboveground biomass C storage (tC ha^{-1}) and total biomass C storage (tC ha^{-1}).

Variables	Forest type	Mean	Standard error	Standard deviation	Minimum	Maximum
GPP	Pine forest	7.67	0.33	2.51	0.11	12.71
	Fir forest	8.25	0.30	2.66	0.07	12.36
	Deciduous forest	7.82	0.10	2.76	0.38	11.93
	Evergreen broad-leaved forest	8.17	0.10	2.28	0.27	12.19
	Total	7.96	0.21	2.59	0.07	12.71
NEP	Pine forest	1.25	0.03	0.56	−1.02	2.19
	Fir forest	1.17	0.03	0.69	−1.04	2.13
	Deciduous forest	1.29	0.07	0.64	−0.86	2.08
	Evergreen broad-leaved forest	1.12	0.08	0.61	−0.69	2.13
	Total	1.2	0.05	0.65	−1.04	2.19
Aboveground biomass C storage	Pine forest	28.02	0.77	16.66	0.17	115.54
	Fir forest	26.68	0.63	17.11	0.09	120.88
	Deciduous forest	25.56	1.72	15.28	0.34	107.68
	Evergreen broad-leaved forest	24.95	2.0	15.32	0.23	90.98
	Total	27.01	1.28	16.78	0.09	120.88
Total biomass C storage	Pine forest	37.24	0.92	20	0.27	136.14
	Fir forest	35.36	0.77	20.78	0.15	141.67
	Deciduous forest	34.26	2.09	18.62	0.61	126.90
	Evergreen broad-leaved forest	33.27	2.45	18.8	0.52	108.90
	Total	35.87	1.56	20.31	0.15	141.67

NEP was greatly determined by GPP ($r = 0.95$), and a strong correlation was found between aboveground and total biomass C storage ($r = 1.0$, $p<0.01$). Aboveground and total biomass C density were more highly correlated to GPP ($r = 0.64$ and 0.69, $p<0.01$) than NEP ($r = 0.47$ and 0.52, $p<0.01$). Stand diameter and height declined significantly ($p<0.01$) as longitude, latitude and elevation increased, but stand density, GPP and NEP increased significantly ($p<0.01$) under the same geographic and topographical conditions.

3.2 *Factors affecting C sequestration in forests*

Site location (longitude, latitude and elevation), climate (frost days, temperature, precipitation and the VPD) and stand structure (density, age, diameter and height) were included in the correlation analysis with GPP, NEP, aboveground and total biomass C density of subtropical forests (Table 2). GPP, NEP, aboveground and total biomass C density were significantly correlated to stand-structure variables ($p<0.01$), except for the relationship between NEP and stand age. Among site location and climate, only longitude and temperature were significantly correlated with GPP, NEP, aboveground and total biomass C density ($p<0.01$) (Table 2). Generally, high GPP and NEP were associated with increased longitude, temperature, stand diameter and height, and high biomass C density was associated with high temperatures, stand density, stand diameter and height (Table 2).

Table 2. Regression parameters for C-sequestration dependent variables (*Y*; GPP, NEP, aboveground biomass and total biomass) against chosen ($p<0.05$) predictors (*X*; longitude, annual average air temperature, stand density, age, average diameter at breast height and average height).

Dependent variable *Y*	Regression* $Y=a+biXi$	Predictor *Xi*	*bi*	Standard error	*t*	Significance
GPP	$r=0.616$; $r^2=0.38$; Standard error=2.04; $p<0.01$.	Constant (*a*)	−16.91	4.64	−3.64	0
		Longitude	0.13	0.04	3.11	0
		Temperature (°C)	0.2	0.04	4.69	0
		Stand density (stems ha^{-1})	0	0	21.58	0
		Age (year)	−0.08	0.01	−11.15	0
		DBH (cm)	0.35	0.03	13.49	0
		H (m)	0.1	0.03	2.97	0
NEP	$r=0.566$; $r^2=0.32$; Standard error=0.53; $p<0.01$.	Constant (*a*)	−4.72	1.21	−3.9	0
		Longitude	0.03	0.01	3.03	0
		Temperature (°C)	0.04	0.01	3.35	0
		Stand density (stems ha^{-1})	0	0	18.84	0
		Age (year)	−0.02	0	−11.8	0
		DBH (cm)	0.1	0	22.8	0
Aboveground biomass C storage	$r=0.656$; $r^2=0.43$; Standard error=12.52; $p<0.01$.	Constant (*a*)	−32.13	4.45	−7.22	0
		Temperature (°C)	0.98	0.25	3.89	0
		Stand density (stems ha^{-1})	0.01	0	19.9	0
		DBH (cm)	0.97	0.16	6.06	0
		H (m)	1.63	0.2	8.32	0
Total biomass C storage	$r=0.671$; $r^2=0.45$; Standard error=14.94; $p<0.01$.	Constant (*a*)	−109.25	34.29	−3.19	0
		Longitude	0.64	0.31	2.06	0.04
		Temperature (°C)	1.20	0.31	3.90	0
		Stand density (stems ha^{-1})	0.02	0.00	20.78	0
		DBH (cm)	1.30	0.19	6.82	0
		H (m)	1.88	0.23	8.06	0

* Method: stepwise multilinear regression.
Criteria: probability-of-F-to-enter ≤0.05, probability-of-F-to-remove ≥1.0

4 DISCUSSION

4.1 *Model simulation and uncertainty*

Owing to high spatial heterogeneity, considerable uncertainty exists in evaluating the role of global and regional C cycles in subtropical forests (Piao et al. 2005). Given the validity of the process-based hybrid-forest model and the availability of primary reliable data from permanent forest sample plots as well as other sources, an overall accurate estimation of C sequestration by forests on a provincial scale can be achieved (Mäkelä et al. 2012). Hybrid models are promising tools by which to integrate processes at different scales, using physiological mechanisms and empirically derived approaches to establish relationships (Peng et al. 2002). Such models require only modest data, few parameters and a means to overcome a lack of mechanistic knowledge through empiricism based on causal thinking (Peng et al. 2002). TRIPLEX1.6 is a robust method that links local and regional scales with site data and stand structure from permanent forest plots, climate records and data derived from literature to estimate C sequestration by forests (Peng et al. 2009, Zhang et al. 2008). A general parameterization and validation scheme for TRIPLEX1.6 has already been established in southeastern and northeastern China (Peng et al. 2009, Zhang et al. 2008). Its specific dynamic parameterization and validation methods are optimized to maximize simulation accuracy for complex regions (Mäkelä et al. 2012, Zhou et al. 2004).

In Hunan Province, the complex landscape includes diverse vegetation types, distinct landforms and high biophysical heterogeneity (Tang et al. 2011). This study adopted scale-specific (i.e., site-, stand- and tree species-specific) dynamic strategies to parameterize and validate TRIPLEX1.6 simulations in Hunan Province, subtropical China, considering knowledge gained from our previous study (Zhao et al. 2013). Nevertheless, possible uncertainties can be present within estimations. For example, uncertain stand-biomass estimations contribute at least half the uncertainty in estimates of land-use change caused by C-source increases (Houghton 2005). Stand biomass, particularly aboveground biomass, is vulnerable to natural and anthropogenic disturbances. TRIPLEX1.6 simulations disregard several variables related to contingency incident factors (i.e., anthropogenic disturbances, harvest, insects, infestations and extreme weather), resulting in diminished accuracy (Peng et al. 2009). Absence of shrub and grass layer sub-models may be another cause for uncertainty in forest C-sequestration estimations in subtropical China (Peng et al. 2009). Moreover, permanent forest-plot data, independent investigations and retrievals may contribute to this uncertainty (Piao et al. 2005).

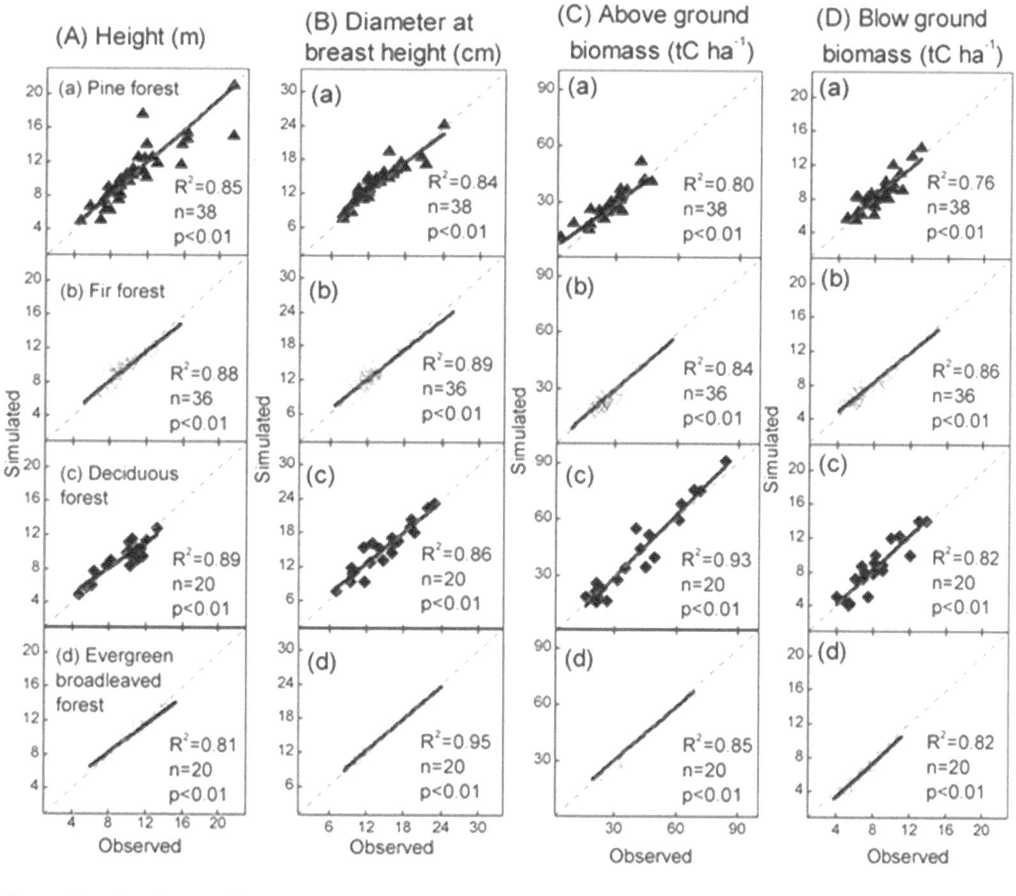

Figure 2. Species-specific validation of simulated and observed stand average height (A), diameter at breast height (B), aboveground (C) and belowground biomass (D) for pine (a), fir (b), deciduous (c) and evergreen broadleaved forests (d) in Hunan Province.

Despite these uncertainties, however, the validation results indicated that TRIPLEX1.6 performed satisfactorily and was able to explain most of the variation in C sequestration by forests in the four subtropical forest types in permanent plots at the stand level throughout Hunan Province ($r2>0.75$, $p<0.01$) (Figure 2). As a result, the simulated output could be extrapolated to evaluate regional C sequestration by forests.

4.2 *Comparison of forest C storage and sequestration rate*

GPP, NEP, aboveground and total stand biomass are indices widely used to estimate C sequestration by forests (Chen et al. 2013, Houghton 2005), which represent the rate and stock of C accumulation over a certain period of time. Average values and ranges of GPP and NEP exhibited higher spatial variation compared to stand-biomass C density, with most plots having high levels of GPP and NEP and low aboveground and total stand-biomass C density. C sinks mainly occurred in areas with low stand-biomass C density. These results are consistent with estimates reported by Piao et al. (2005), who used forest inventory and satellite data (Piao et al. 2005). This result suggests that afforestation and reforestation play key roles in climate-change mitigation (Zhou et al. 2013).

Although our average NEP estimates for Hunan Province forests (1.20 t C ha-1 yr-1) were considerably lower than those of Yu et al. (2014) (3.41 t C ha^{-1} yr^{-1}), for the same timeframe, they fell within the lower range of another estimate (1.15-2.27 t C ha-1 yr^{-1}) between 1982 and 2004 (Tang et al. 2011). Lower NEP estimates from this study are likely related to tree age and wet N deposition (Liu et al. 2013, Stephenson et al. 2014, Tan et al. 2011, Zhang et al. 2014). Hunan Province forests are recovering from several past disturbances. Accordingly, the lower estimates found in this study may be the result not only of younger stand samples, including an inventory of 1286 permanent forest-sample plots that ranged in age from 4 to 68 years, but of sample plots that mostly (90%) consisted of young-to-middle-aged stands (age <30; Figure 1) with lower wet

N deposition (18 kg N ha^{-1} yr^{-1}) (Chen & Mulder, 2007) used for this study. Yu et al. (2014) studied undisturbed old-growth samples (age = 110 ± 60 years), including five flux towers in Jiangxi, Yunnan and Guangdong provinces and used higher wet N deposition (25 kg N ha-1 yr^{-1}) in their subtropical-forest estimation. Thus, higher atmospheric wet N deposition might have considerable potential to increase average forest NEP in Hunan Province.

Stand-biomass C-density estimations were 37.24 t C ha^{-1} for pine forests, 35.36 t C ha^{-1} for fir forests, 34.26 t C ha^{-1} for deciduous broadleaved forests and 35.87 t C ha-1 provincial average in Hunan Province. These are consistent with previous estimations (25.77-45.31 t C ha_{-1}) derived from inventory datasets, remote-sensing data or field investigations over the past two decades in China (Fang et al. 2001, Piao et al. 2005), with the exception of certain evergreen broadleaved forest underestimations, such as Zhejiang (33.27 t C ha^{-1} versus 44.60 t C ha^{-1}) (Zhang et al. 2007). The lower estimates of evergreen broadleaved forest-biomass C density from this study reflect spatial heterogeneities related to site, climate and stand structure, likely related to past disturbances, tree-species composition, management practices and site conditions (Piao et al. 2005).

The Hunan Province forests mainly consisted of young plantations and middle-aged natural forests, both in their fast-growing stages (average stand age, 18 years), compared to mature forests with more economic (20 to 25 years) and ecological (25 to 30 years) applications. In 2009, the forests had a higher average GPP (7.96 t C ha^{-1} yr^{-1}) and lower average total biomass density (35.87 t C ha^{-1}) compared to national and global forest averages. The total forests in Hunan Province accounted for about 9.08% of all forested land area in China.

Moreover, total forest-biomass C stock was 0.45 Pg C, accounting for 7.76% of total biomass C storage in China (5.79 Pg C) (Piao et al. 2005). Therefore, Hunan Province forests play an important role in both regional and national C budgets.

4.3 *Identification of controlling factors*

The results of this study showed that site location and climate factors affected C-sequestration rates more strongly than they affected C storage, while stand structure affected C storage more strongly than they affected C-sequestration rates. This finding is consistent with other research results on regional and local scales (Yu et al. 2013). In addition to the effects of stand age on GPP and NEP, the main factors that controlled GPP and NEP included site location (longitude), climate (temperature) and stand structure (stand density, diameter and height). The factors that controlled aboveground and total biomass C storage were stand structure (stand density, age, diameter and height). Stand age (not C storage) structure (e.g., the fast-growing young- and middle-aged stages) was the key driver of C-sequestration rates at the provincial level, while longitude significantly affected the spatial heterogeneity of forest-biomass C density. This result contradicts previous research that reported that site location (longitude and latitude) and climate (temperature and precipitation) determine spatial patterns of C balance in China (Yu et al. 2013, Zhao & Zhou 2006).

5 CONCLUSIONS

Owing to the complexity of spatial heterogeneity, considerable uncertainties arise when estimating C sequestration by forests in subtropical China. The integration of available data sources (such as permanent forest plots and climate data) and the application of a process-based hybrid model (such as TRIPLEX1.6) offer an approach by which to increase accuracy, while reducing quantification uncertainties inherent to C sequestration by forests investigations. TRIPLEX1.6 simulations estimated C sequestration at 0.45 Pg C for the total stand biomass of all forests in Hunan Province in 2009, with 35.84 t C ha-1 average total stand-biomass density, 26.98 t C ha^{-1} aboveground biomass, 7.96 t C ha^{-1} yr^{-1} GPP and 1.203 t C ha^{-1} yr^{-1} NEP. Changes in site location, climate and stand structure resulted in variable forest C-sequestration capacity. Stand age was the key driver of C-sequestration rate provincial-scale dynamics, while longitude was significantly affected the spatial heterogeneity of forest-biomass C density. Forest management practices that maintain and promote larger trees and denser stands could potentially increase regional C-sequestration rates and large C sinks in subtropical forests in China.

ACKNOWLEDGEMENTS

This study received grants from the National Key Research and Development Program (973 Program) (2013CB956602 and 2010CB833501), the Program of Basic Platform for National Ecosystem Observation and Research Network, and the Furong Scholar Program.

REFERENCES

Arlinghaus, A., Lombardi, D.A., Willetts, J.L., Folkard, S., Christiani, D.C. 2012. A structural equation modeling approach to fatigue-related risk factors for occupational injury. *Am. J. Epidemiol.* 176:597–607.

Bossel, H. 1996. TREEDYN3 forest simulation model. *Ecol. Model.* 90:187–227.

Chen, Q., Xu, W., Li, S., Fu, S., Yan, J., 2013. Aboveground biomass and corresponding carbon sequestration ability of four major forest types in south China. *Chinese Sci. Bull.* 58:1551–1557.

Chen, X.Y., Mulder, J., 2007. Atmospheric deposition of nitrogen at five subtropical forested sites in south China. *Sci. Total Environ.* 378:317–330.

Falge, E., Baldocchi, D., Olson, R., Anthoni, P., Aubinet, M., Bernhofer, C., Burba, G., Ceulemans, R., Clement, R., Dolman, H., Granier, A., Gross, P., Grünwald, T., Hollinger, D., Jensen, N.-O., Katul, G., Keronen, P., Kowalski, A., Lai, C.T., Law, B.E., Meyers, T., Moncrieff, J., Moors, E., Munger, J.W., Pilegaard, K., Rannik, U.l., Rebmann, C., Suyker, A., Tenhunen, J., Tu, K.,

Verma, S., Vesala, T., Wilson, K., Wofsy, S. 2001. Gap filling strategies for defensible annual sums of net ecosystem exchange. *Agric. For. Meteorol.* 107:43–69.
Fang, J., Chen, A., Peng, C., Zhao, S., Ci, L. 2001. Changes in forest biomass carbon storage in China between 1949 and 1998. *Science* 292:2320–2322.
Food Agricultural Organization (FAO), 2010. Global forest resources assessment 2010. *Forestry Paper*, 163, Rome.
Houghton, R.A. 2005. Aboveground forest biomass and the global carbon balance. *Glob. Chang. Biol.* 11, 945–958.
Lin, D., Lai, J., Muller-Landau, H.C., Mi, X., Ma, K. 2012. Topographic variation in aboveground biomass in a subtropical evergreen broad-leaved forest in China. PloS ONE 7, e48244.
Liu, Y., Yu, G., Wang, Q., Zhang, Y. 2013. How temperature, precipitation and stand age control the biomass carbon density of global mature forests. *Glob. Ecol. Biogeogr.* 23:323–333.
Mäkelä, A., Río, M.d., Hynynen, J., Hawkins, M.J., Reyer, C., Soares, P., van Oijen, M., Tomé, M. 2012. Using stand-scale forest models for estimating indicators of sustainable forest management. *For. Ecol. Manage.* 285:164–178.
Pan, Y., Birdsey, R.A., Fang, J., Houghton, R., Kauppi, P.E., Kurz, W.A., Phillips, O.L., Shvidenko, A., Lewis, S.L., Canadell, J.G., Ciais, P., Jackson, R.B., Pacala, S.W., McGuire, A.D., Piao, S., Rautiainen, A., Sitch, S., Hayes, D. 2011. A large and persistent carbon sink in the World's forests. *Science* 333:988–993.
Parton, W.J., Scurlock, J.M.O., Ojima, D.S., Gilmanov, T.G., Scholes, R.J., Schimel, D.S., Kirchner, T., Menaut, J.C., Seastedt, T., Garcia Moya, E., Kamnalrut, A., Kinyamario, J.I., 1993. Observations and modeling of biomass and soil organic matter dynamics for the grassland biome worldwide. *Glob. Biogeochem. Cycles* 7:785–809.
Peng, C., Liu, J., Dang, Q., Apps, M.J., Jiang, H., 2002. TRIPLEX: A generic hybrid model for predicting forest growth and carbon and nitrogen dynamics. *Ecol. Model.* 153:109–130.
Peng, C.H., Zhou, X.L., Zhao, S.Q., Wang, X.P., Zhu, B., Piao, S.L., Fang, J.Y. 2009. Quantifying the response of forest carbon balance to future climate change in northeastern China: Model validation and prediction. *Glob. Planet. Chang.* 66:179–194.
Piao, S., Ciais, P., Lomas, M., Beer, C., Liu, H., Fang, J., Friedlingstein, P., Huang, Y., Muraoka, H., Son, Y., Woodward, I., 2011. Contribution of climate change and rising CO2 to terrestrial carbon balance in East Asia: A multi-model analysis. *Glob. Planet. Chang.* 75:133–142.
Piao, S., Fang, J., Zhu, B., Tan, K., 2005. Forest biomass carbon stocks in China over the past 2 decades: Estimation based on integrated inventory and satellite data. *J. Geophys. Res.: Biogeosci.* 110:C01006.
Song, X., Yuan, H., Kimberley, M.O., Jiang, H., Zhou, G., Wang, H., 2013. Soil CO_2 flux dynamics in the two main plantation forest types in subtropical China. *Sci. Total. Environ.* 444:363–368.
Stephenson, N.L., Das, A.J., Condit, R., Russo, S.E., Baker, P.J., Beckman, N.G., Coomes, D.A., Lines, E.R., Morris, W.K., Ruger, N., Alvarez, E., Blundo, C., Bunyavejchewin, S., Chuyong, G., Davies, S.J., Duque, A., Ewango, C.N., Flores, O., Franklin, J.F., Grau, H.R., Hao, Z., Harmon, M. E., Hubbell, S.P., Kenfack, D., Lin, Y., Makana, J.R., Malizia, A., Malizia, L.R., Pabst, R.J., Pongpattananurak, N., Su, S.H., Sun, I.F., Tan, S., Thomas, D., van Mantgem, P.J., Wang, X., Wiser, S.K., Zavala, M.A., 2014. Rate of tree carbon accumulation increases continuously with tree size. *Nature* 507:90–93.
Tang, X., Wang, Y., Zhou, G., Zhang, D., Liu, S., Liu, S., Zhang, Q., Liu, J., Yan, J., 2011. Different patterns of ecosystem carbon accumulation between a young and an old-growth subtropical forest in southern China. *Plant Ecol.* 212:1385–1395.
Wang, Z., Grant, R.F., Arain, M.A., Bernier, P.Y., Chen, B., Chen, J.M., Govind, A., Guindon, L., Kurz, W.A., Peng, C., Price, D.T., Stinson, G., Sun, J., Trofymowe, J.A., Yeluripati, J. 2013. Incorporating weather sensitivity in inventory-based estimates of boreal forest productivity: A meta-analysis of process model results. *Ecol. Model.* 260:25–35.
Xiang, W., Liu, S., Deng, X., Shen, A., Lei, X., Tian, D., Zhao, M., Peng, C., 2011. General allometric equations and biomass allocation of Pinus massoniana trees on a regional scale in southern China. *Ecol. Res.* 26:697–711.
Yin, G., Zhang, Y., Sun, Y., Wang, T., Zeng, Z., Piao, S., 2015. MODIS based estimation of forest aboveground biomass in China. *PLoS ONE* 10, e0130143.
Yu, G., Chen, Z., Piao, S., Peng, C., Ciais, P., Wang, Q., Li, X., Zhu, X., 2014. High carbon dioxide uptake by subtropical forest ecosystems in the east Asian monsoon region. *Proc. Natl. Acad. Sci. U. S. A.* 111: 4910–4915.
Yu, G., Zhang, L., Sun, X., Fu, Y., Wen, X., Wang, Q., Li, S., Ren, C., Song, X.I.A., Liu, Y., Han, S., Yan, J., 2008. Environmental controls over carbon exchange of three forest ecosystems in eastern China. *Glob. Chang. Biol.* 14:2555–2571.

Yu, G.R., Zhu, X.J., Fu, Y.L., He, H.L., Wang, Q.F., Wen, X.F., Li, X.R., Zhang, L.M., Zhang, L., Su, W., Li, S.G., Sun, X.M., Zhang, Y.P., Zhang, J.H., Yan, J.H., Wang, H.M., Zhou, G.S., Jia, B. R., Xiang, W.H., Li, Y.N., Zhao, L., Wang, Y.F., Shi, P.L., Chen, S.P., Xin, X.P., Zhao, F.H., Wang, Y.Y., Tong, C.L., 2013. Spatial patterns and climate drivers of carbon fluxes in terrestrial ecosystems of China. *Glob. Chang. Biol.* 19:798–810.

Zhang, H., Song, T., Wang, K., Wang, G., Liao, J., Xu, G., Zeng, F., 2015. Biogeographical patterns of forest biomass allocation vary by climate, soil and forest characteristics in China. *Environ. Res. Lett.* 10:044014.

Zhang, J., Chu, Z., Ge, Y., Zhou, X., Jiang, H., Chang, J., Peng, C., Zheng, J., Jiang, B., Zhu, J., Yu, S., 2008. TRIPLEX model testing and application for predicting forest growth and biomass production in the subtropical forest zone of China's Zhejiang Province. *Ecol. Model.* 219:264–275.

Zhang, J., Ge, Y., Chang, J., Jiang, B., Jiang, H., Peng, C., Zhu, J., Yuan, W., Qi, L., Yu, S., 2007. Carbon storage by ecological service forests in Zhejiang Province, subtropical China. *For. Ecol. Manage.* 245: 64–75.

Zhao, M., Zhou, G., 2006. Carbon storage of forest vegetation in China and its relationship with climatic factors. *Clim. Chang.* 74:175–189.

Zhao, M., Xiang, W., Deng, X., Tian, D., Huang, Z., Zhou, X., Yu, G., He, H., Peng, C., 2013a. Application of TRIPLEX model for predicting cunninghamia lanceolata and Pinus massoniana forest stand production in hunan province, southern China. *Ecol. Model.* 250:58–71.

The Role of Agriculture in Climate Change Mitigation – Pawłowski, Litwińczuk & Zhou (eds)

Effects of neighborhood tree species diversity on soil organic carbon and labile carbon in subtropical forest

S. Sun, X. Song, Y. He, Q. Qian & Y. Yao
Faculty of Life Science and Technology, Central South University of Forestry and Technology, Changsha, Hunan, China

P. Lei, W. Xiang & S. Ouyang
Huitong National Field Station for Scientific Observation and Research of Chinese Fir Plantation Ecosystem in Hunan Province, Huitong, China

1 INTRODUCTION

The soil carbon pool is the largest carbon pool in the terrestrial ecosystem carbon pools. As an important source and sink of atmospheric CO_2 concentration, the soil carbon cycle becomes an important factor affecting the global carbon cycle. The SOC plays a key role in the function of forest ecosystems, since it offers energy and substance for microbial metastasis and improves biodiversity (Loveland & Webb 2003), At present, the global soil carbon pool is about 2500 Gt, and the soil organic carbon pool account for nearly 1550 Gt (Lal 2004). The dynamic changes of soil organic carbon may affect the climate in the short period (Schlesinger 1982). The variety of soil carbon storage has a close relationship with the plant nutrition and soil (Shen & Cao 1999), for example, fertilization has great significance on soil C availability (Shen & Cao 1998).

And the soil carbon with different forms exists in the soil, such as soil organic carbon (SOC), mineralizable carbon (MLC), soil microbial biomass carbon (MicC), easily-oxidized carbon (EOC) and so on. Among them, the MLC is mainly source of input of CO2 from soil carbon pools into atmospheric carbon pools and plays a crucial role in global climate change (Alexander 1961). MOC performed an affordable active organic matter measurements in soil health constructions (Motavalli et al. 1994, Wade et al. 2018). And it depends largely on the decomposition of litterfall. The strength and magnitude of CO_2 released by soil carbon mineralization can be served as an target to test the decomposition ratio of soil organic carbon (Moebius-Clune et al. 2016). The MicC is an important indicator of soil organic matter decomposition, although it usually explains only 1 to 4 percent of total organic carbon (Jenkinson 1981). Microbial biomass and mineral sizable C may reflect soil productivity (Sparling 1992, Hassink 1993, Chen & Wen 1998), and has significantly role in nutrient cycling and transformations, and can store much valid nutrients (Kandeler et al.1999, Ge et al. 2013, Spedding et al. 2004). The EOC can effectively reflect the soil quality, which was the most active part of soil carbon and more sensitive than other indicators in soil (Weil et al. 2003; Jiang et al. 2006). The DOC is a major component of dissolved organic matter, which is a very active chemical component in the soil and participated in the process of carbon and nitrogen cycle and biochemical cycle, Simultaneously it is an immediate source of plant nutrients (Chen & Xu 2008, Blair et al 1995). Due to the High turnover rate, both DOC and MicC are considered as early indicators of C storage in soil (Scaglia & Adani 2009).

In this study we aimed to investigate influences of the tree species richness at a fine scale on SOC and stability in natural regenerated forests. More specifically, the research objectives were to test hypotheses: (i) Overall SOC increases with tree species richness; (ii) this increased carbon sequestration were attributed to increased aboveground and belowground litter inputs; and (iii) labile carbon increased with tree species diversity as the increased substrate quality.

2 MATERIALS AND METHODS

2.1 *Site description and experimental design*

The experiment site was located at Dashanchong Forest Park in Changsha County, Hunan Province (28°23′58″-28°24′58″N, 113°17′46″-113°19′08″E). The range of altitude is 55-217.1 m above sea level. Dashanchong Forest Park has good vegetation and rich tree species, which can represent typical subtropical vegetation. The climate in the area belongs to subtropical monsoon climate. The soil is red soil developed from slate and shale. The zonal vegetation is evergreen broad-leaved forest. One permanent site of 1 ha (100 m × 100 m) were established in 2013 and all the trees with DBH>10 cm were marked with GPS and All tree stems with a DBH ≥ 4 cm were recorded for the diameter at breast height (DBH), total height (H), and species name. In this study, four typical subtropical tree species, including two early succession trees of *Pinus massoniana* and *Choerospondias axillaris* and two late succession trees of *Cyclobalanopsis glauca* and *Lithocarpus glaber*, were selected as research objects and the basic information of each tree is shown in the Table 1. A variety of forest tree cluster consisting 3 to 4 trees in gradient of tree species richness from 1, 2, 3 to 4 tree species. In total there were 15 combinations, including four combinations of 1 species monoculture, six combinations of 2 species, 4 combinations of 3 species and 1 combination of 4 species. Considering the cumulative and long term carbon formation, here only the tree with DBH bigger than 10 cm were selected as effective tree selecting criteria. Each combination were replicated three times. The selected tree clusters consisting 3 to 4 trees in gradient of tree species richness were layout in Figure 1.

2.2 *Soil sampling and soil analysis methods*

Soil samples were collected in the central point of each selected tree clusters with soil auger of 15 cm in diameter and sliced to four layers (0–15, 5-30, 30-45 and 45-60 cm) in April, 2017. In the laboratory, the soil core was sieved through a 2-mm mesh sieve to collect the soil samples for soil analysis, and the unsieved part were washed with tap water and all the fine root were collected with 0.5 mm sieve. Before measurement, the soil sample was divided into two parts. One was homogenized and then stored at 4 °C until analysis for microbial biomass. The other part of soil was dried at room temperature and then sieved through 0.25 mm sieve. Soil pH value was determined on a 1:2 (soil: water) mixed suspension with an electrode. SOC was determined by applying the wet combustion method using oxidization of potassium bichromate (Walkley–Black method) and then titrated with 0.5 N ferrous ammonium sulfate solution by using diphenyl-amine indicator.

MLC was estimated by a bioassay technique based on NaOH absorption method cultivated for 3 days of C mineralization under favorable laboratory conditions. EOC was determined by 333 mM $KMnO_4$. TN was measured using the Semimicro–Kjeldahl method digested with a mixture of H_2SO_4, K_2SO_4, $CuSO_4$ and Se (Institute of Soil Science, Chinese Academy of Science 1978). Soil microbial carbon (Cmic) and microbial nitrogen (Nmic) was determined using the chloroform fumigation extraction method. Around 10g fresh soils were weighted

Table 1. Floristic composition and community structure for top four dominate species and whole stand with diameter at breast height (DBH) larger than 4 cm in *L. glaber - C. glauca* forest.

Species	Density (stem·ha-1)	DBH (cm)	Height (m)	Basal area (m2·ha-1)	Importance value
Lithocarpus glaber	586	10.4(4.0-37.1)	. 6(2.2-19.5)	6.70	25.93
Cyclobalanopsis glauca	164	12.8(4.0-34.6)	10.4(3.8-20.0)	3.01	9.90
Pinus massoniana	123	18.0(7.0-32.2)	14.2(1.3-20.0)	3.46	9.77
Choerospondias axillaris	83	19.3(4.0-46.8)	13.5(1.8-20.2)	2.85	7.91
total	956	15.1(4.0-46.8)	11.9(1.2-21.0)	16.02	53.51

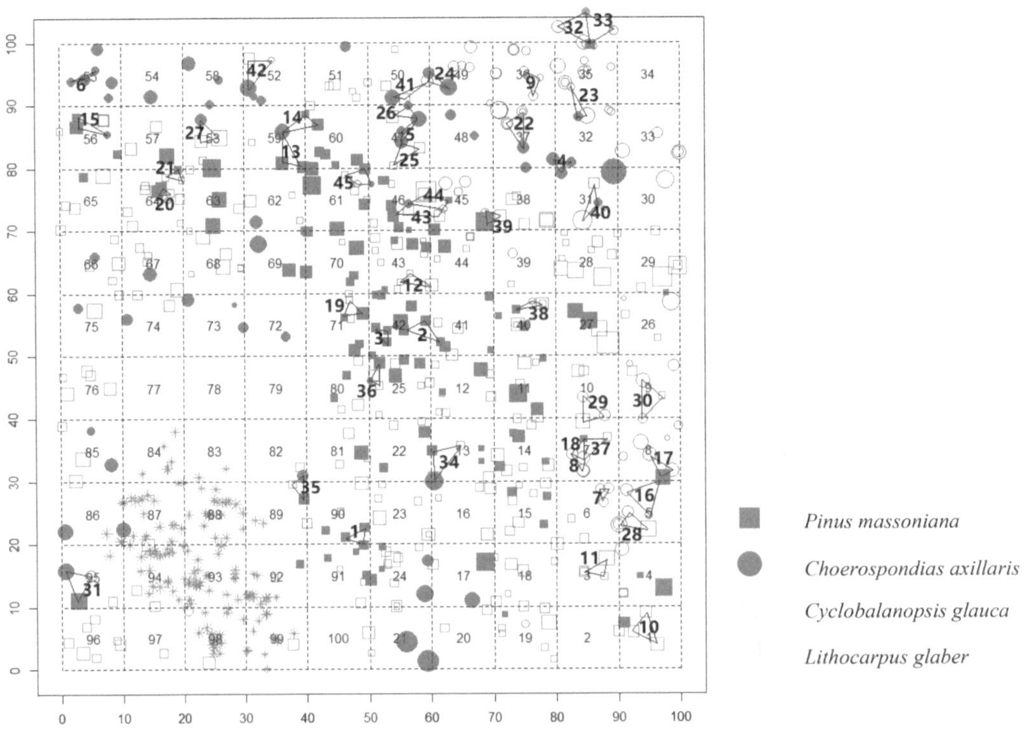

Figure 1. Layout of tree clusters in gradient of tree species richness.

and fumigated in vacuum desiccators filled with alcohol-free chloroform at 25°C for 24 h. At the same time, 10 gram unfumigated soil were also prepared. Then both fumigate and unfumigated soils were extracted with 0.5 M K_2SO_4. Total organic carbon and nitrogen in fumigated and unfumigated extracts was measured with Multi N/C 3100 TOC-TN analyzer (Analytic Jena, Germany). The Cmic and Nmic were calculated as the concentration of organic C in fumigated samples subtracted by that in unfumigated (control) samples with a conversion factor of 0.45 for microbial C. The K2SO4-extractable C extracted from unfumigated soil was used as another measure of labile soil organic C (DOC).

2.3 *Soil sampling and soil analysis methods*

Soil samples were collected in the central point of each selected tree clusters with soil auger of 15 cm in diameter and sliced to four layers (0–15, 5-30, 30-45 and 45-60 cm) in April, 2017. In the laboratory, the soil core was sieved through a 2-mm mesh sieve to collect the soil samples for soil analysis, and the unsieved part were washed with tap water and all the fine root were collected with 0.5 mm sieve. Before measurement, the soil sample was divided into two parts. One was homogenized and then stored at 4°C until analysis for microbial biomass. The other part of soil was dried at room temperature and then sieved through 0.25 mm sieve. Soil pH value was determined on a 1:2 (soil: water) mixed suspension with an electrode. SOC was determined by applying the wet combustion method using oxidization of potassium bichromate (Walkley–Black method) and then titrated with 0.5 N ferrous ammonium sulfate solution by using diphenyl-amine indicator. MLC was estimated by a bioassay technique based on NaOH absorption method cultivated for 3 days of C mineralization under favorable laboratory conditions. EOC was determined by 333 mM $KMnO_4$. TN was measured using the Semimicro–Kjeldahl method digested with a mixture of H_2SO_4, K_2SO_4, $CuSO_4$ and Se (Institute of Soil Science, Chinese Academy of Science

1978). Soil microbial carbon (Cmic) and microbial nitrogen (Nmic) was determined using the chloroform fumigation extraction method. Around 10g fresh soils were weighted and fumigated in vacuum desiccators filled with alcohol-free chloroform at 25 °C for 24 h. At the same time, 10 gram unfumigated soil were also prepared. Then both fumigate and unfumigated soils were extracted with 0.5 M K_2SO_4. Total organic carbon and nitrogen in fumigated and unfumigated extracts was measured with Multi N/C 3100 TOC-TN analyzer (Analytic Jena, Germany). The Cmic and Nmic were calculated as the concentration of organic C in fumigated samples subtracted by that in unfumigated (control) samples with a conversion factor of 0.45 for microbial C. The K_2SO_4-extractable C extracted from unfumigated soil was used as another measure of labile soil organic C (DOC).

2.4 *Fine root biomass and litterfall measurement*

After sieved with 2 mm-mesh sieve, the residues of the soil cores were soaked in the water and gently squeezed to small pieces. The root would float and poured the water on 0.5 mm-mesh sieved. And all the fine roots were collected. Then the fine roots were separated into grass and woody species. The sorted live roots were over-dried at 80°C to constant weight to estimate for fine root biomass. For aboveground litterfall measurement, one round litter trap of 1 m^2 was placed in the central point of each tree species combination cluster in February 2017. The litterfall obtained with this litter trap were extracted monthly thereafter until February. The litterfall were sorted to species and components, including leaves, branches, trigs, fruits and residues.

2.5 *Data analysis*

A single-factor analysis of variance (ANOVA) was used to conduct different comparisons among the four levels of tree species richness for each specific soil layers at $P<0.05$, and the means were separated using least significant difference (LSD). Pearson correlation analysis was performed between SOC, MLC, EOC, DOC MicC and soil pH, moisture, litterfall and fine root biomass. All analyses were conducted with the statistical software R (v.3.3.2) (R Development Core Team 2016).

3 RESULTS

3.1 *Soil organic carbon and labile carbon fractions*

The concentrations of SOC decreased with increasing soil depth. When comparing the SOC among different species diversity level, the SOC concentration were significantly higher in mixed combination of tree clusters than that in one specie monocultures in top soil layer (0-15 cm), while the differences among then in deeper soil layers (15-30, 30-45 and 45-60 cm) were not significant (Figure 2). In 0-15 cm soil depth, the 3 tree species combinations showed the highest SOC compared to the 2 and 4 tree species combinations, but the differences were not significant.

In Figure 3, it shows the content of MLC and EOC between different tree species richness under different soil layers. In the 0-15cm soil, the content of MLC of richness 2 was higher than other richness; in 15-30 cm, the content of richness 3 of MLC is higher than richness 2 and 4; in 30-45 cm, richness 4 had the least content of MLC, and content of MLC of richness 1 was the highest; in the soil layer of 45-60cm, richness 1 had higher content of MLC, and richness 4 had lower mineralizable content, and the trend is similar with the third layer. In 0~15 cm, the content of EOC had a litter difference; in 15~30cm, the content of EOC in richness 3 was higher than other richness, the lowest was in richness 4; in 30~45cm, the content of EOC among richness 4 was higher than other richness, the lowest was in richness 1 and 45~60 cm, the content of EOC in richness 1 was higher than others, the lowest was in richness 4.

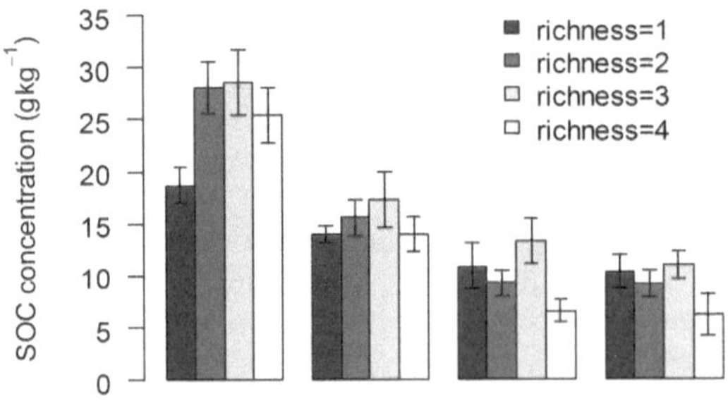

Figure 2. Effects of tree species richness on SOC in 0-15, 15-30, 30-45 and 45-60 cm soil depth. Error bars indicate standard errors. Different letters indicate significant differences between different species richness within the same soil profile ($P < 0.05$).

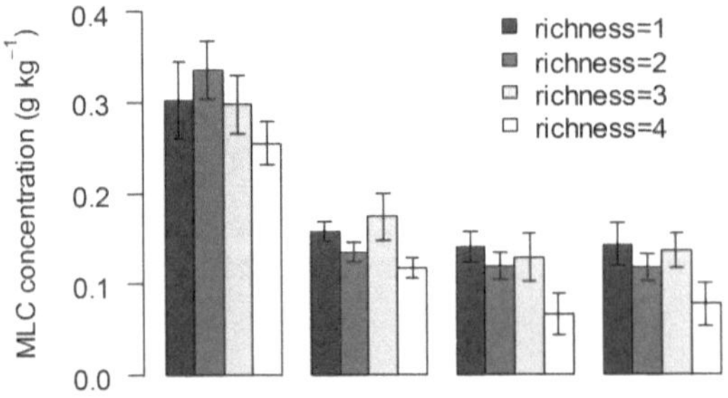

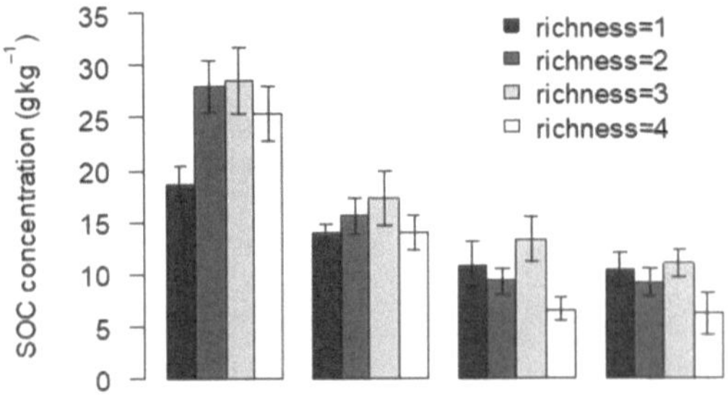

Figure 3. Effects of tree species richness on MLC and EOC in 0-15, 15-30, 30-45 and 45-60 cm soil depth. Error bars indicate standard errors. Different letters indicate significant differences between different species richness within the same soil profile ($P < 0.05$).

Figure 4 shows the distribution of MicC concentration and DOC concentration, the MicC concentration decreased as the increasing tree species richness, and the richness 4 was lowest. In the 0-30cm, the content of DOC of the combination of richness 4 was higher than other richness, and the richness 2 had the lowest content.

3.2 *Ratios of labile carbon to soil organic carbon*

In Figure 5. the ratio of MLC to SOC of richness 1 was the same as richness 2, and the ratio of richness 4 is the lowest among four different tree species richness, the lowest ratio of EOC to SOC was richness 2, and richness 4 had the highest ratio of EOC to SOC. The difference among this four tree species richness is not very obvious, particularity between the combination of two species and three species. From the chart, the ratio of DOC to SOC between different tree species richness, the richness 3 was lower than others, and richness 1 was higher. The ratio of MicC to SOC in richness 1 was higher than others, as the increasing of tree species richness, the ratio of MicC to SOC decreased.

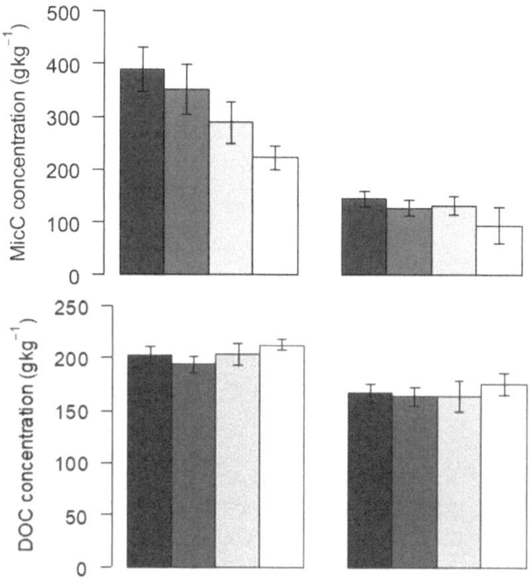

Figure 4. Effects of tree species richness on MicC and DOC in 0-15 and 15-30 cm soil depth. Error bars indicate standard errors. Different letters indicate significant differences between different species richness within the same soil profile (P < 0.05).

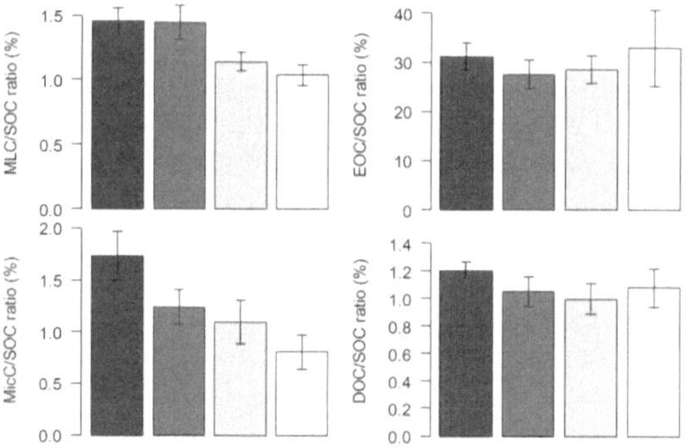

Figure 5. Effects of tree species richness on MLC to SOC ratio, EOC to SOC ratio, MicC to SOC ratio and DOC to SOC ratio. Error bars indicate standard errors. Different letters indicate significant differences between different species richness (P < 0.05).

3.3 *Aboveground litterfall and belowground fine root biomass*

As the Figure 6, which shows the biomass of aboveground litterfall and belowground fine root, among litterfall biomass, the value of richness 4 is significantly lower than other richness and in other richness they have not obvious difference, the richness 1 is a litter higher than others; among fine root biomass, the value in richness 2 is higher than other richness and the richness 3 is lowest.

3.4 *Correlations between SOC, labile carbon fraction and soil properties*

From Figure 6, MLC, EOC, DOC and MicC had a significantly positive relationship with each other, except for DOC and MicC, which were also performed significantly or extreme significantly relationships with SOC and TN, It can be seen from this, under the different tree

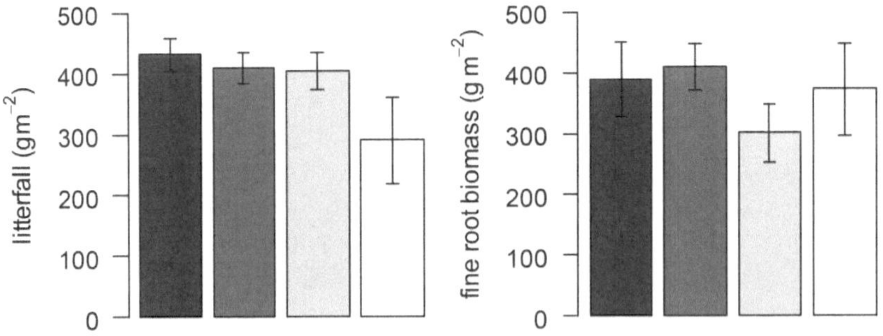

Figure 6. Effects of tree species richness on annual litterfall and fine root biomass. Error bars indicate standard errors. Different letters indicate significant differences between different species richness ($P < 0.05$).

species richness, the value of soil active organic carbon largely depends on the total organic carbon content, which directly participated in the process of soil biochemical conversion and achieved mutual conversion. at the same time, the SOC, MLC, EOC, DOC and MicC also had a significantly positive relationship with root biomass and moisture, but showed extremely significant negative correlation with pH.

4 DISCUSSION

4.1 *Soil organic carbon and labile carbon fractions*

Under this four different tree species richness, the content of SOC decreases with increasing soil depth, this may due to the fact that litter on plants is mainly concentrated on the soil surface, the density of plant roots decreases with the deepening of soil layers. In 0-15cm, the concentrations of SOC were higher in the mixed combinations than that in monoculture and the content of SOC in richness 3 was slightly higher than other tree species richness, and except for the 0-15cm soil layer, the content in richness 4 was lower than other tree species richness, this may be the quantity of litterfall in richness 4 was lower than other richness, and when mixed with multiple tree species ($n>3$), the structure of forest is more complex and exist intensely competition.

MLC is important for biochemical process, which can directly influence the cycle of nutrient and the emissions of greenhouse gas (Jiang et al 2006). In general, the content of MLC in richness 4 was the lowest under each soil layer, which may result from the low value of litterfall and root biomass, in 0-15cm, the content of MLC in richness 2 was higher than other richness, this may be the concentration of litterfall (Figure 6) was a little higher and the fine root biomass was higher than others, it can provide more substantial for microorganism and to some extent this may increase the quantity and activity of microorganism, thus accelerating the process of nutrient cycle and then improving the content of MLC.

EOC is an active part of soil organic carbon pool, which reflects the stability of soil carbon pool (Haynes 2005). In this study, the content of EOC in richness 4, except to 0-15cm, was lower than other tree species richness, which was similar to the distribution of SOC in different tree species richness. This phenomenon may be caused by:(1) Due to the differences in litterfall and root exudates under the forest in different tree species richness, the distribution of the content of EOC varies; (2) When multiple tree species are mixed ($n>3$), in each richness, the growth status and spatial distribution pattern are more complex than simple tree species, resulting in different litter, root exudates, and formed microclimate environments; (3) There are differences in erosion and leaching intensity of surface soil.

DOC is a more active component in soil organic carbon pool, which is the significant energy source of microorganisms, therefore it can directly affect the quantity and activity of microorganism and play an important role in nutrient cycle of ecosystem. From Figure 4, as the increasing of soil layer, the content of DOC decreased, and the content in richness 4 was

Table 2. Pearson correlations between SOC and labile carbon fraction and soil pH, moisture, litterfall and fine root.

	SOC	MLC	EOC	DOC	MicC	MLC/SOC	EOC/SOC	DOC/SOC	MicC/SOCC
SOC	1***					-0.342***	-0.424***	-0.762***	-0.326**
MLC	0.710***	1***				0.207**	-0.170*	-0.305**	0.090ns
EOC	0.510***	0.521***	1***			0.033ns	0.345***	-0.278**	0.061ns
DOC	0.228*	0.395***	0.368***	1***		0.207ns	0.197ns	0.246*	0.324**
MicC	0.244*	0.465***	0.354***	0.435***	1***	0.257*	0.168ns	-0.001ns	0.726***
pH	-0.479***	-0.500***	-0.326***	-0.533***	-0.368***	-0.024ns	0.142ns	0.016ns	-0.241*
Moisture	0.158*	0.232**	0.186*	0.249*	0.296**	0.038ns	0.004ns	0.119ns	0.210ns
TN	0.517***	0.371***	0.244**	0.179ns	0.172ns	-0.163*	-0.185*	-0.358*	-0.153ns
litterfall	0.095^{ns}	0.005ns	-0.008ns	-0.248ns	0.293*	-0.130ns	0.002ns	-0.090ns	0.134ns
Root biomass	0.419***	0.514***	0.325***	0.309*	0.556***	0.034ns	-0.100ns	-0.071ns	0.361***

slightly higher than other tree species richness, the difference was not obvious. By contrast, the quantity of litterfall was obviously lower than others, so the content of DOC has a negatively relationship with the quantity of litterfall, this trend was different with previously studies, it may result from the use of microbial. Simultaneously, the microclimate under the forest may causes differences in soil physical and chemical properties and then affect the distribution of DOC content. Overall, the MLC, EOC and DOC fluctuated along the gradient of species richness and not have a directly relationship with species richness.

MicC is the most active part of soil carbon pool, which can accurately reflect the quantity and activity of microorganism. From Figure 4, as the increasing of soil layer, the content of MICC decreased, which also has a negatively relationship with tree species richness, this phenomenon was obviously different with previous studies, which may arise from the completions between different tree species became more intensely, as the increasing of tree species richness. And in Figure 6, as the tree species richness increasing, the content of litterfall shows a decreasing trend. Therefore, under richness 1, high litter content can provide sufficient decomposition substrate for soil microorganisms and as this experiment was sampled in the fall, when the soil temperature and humidity are appropriate for microorganism survive, it may strengthen the activity of microorganism and thus improve the cycle of soil nutrients.

4.2 *Ratios of labile carbon to soil organic carbon*

The ratio of soil active organic carbon to soil organic carbon can reflect the stability of soil organic carbon pool. Under different tree species richness, due to the difference between root exudates, fine root biomass and the content of SOC, it can cause the different ratio. In this study, the ratio of MLC to SOC and ratio of MicC to SOC decreased with neighborhood species diversity, in richness 1, besides the ratio of EOC to SOC lower than richness 4, the other ratios all higher than other richness, it can be seen that the activity of soil organic carbon was the highest, the stability was the lowest in richness 1 and it can most susceptible to external factors. Therefore, increasing tree species diversity may increase the SOC stability and thus enhance the soil carbon sequestration potential.

4.3 *Factors affecting soil organic carbon and labile carbon factions*

From Figure 6, MLC, EOC, DOC and MicC had a significantly positive relationship with each other, except for DOC and MicC, which were also performed significantly or extreme significantly relationships with SOC and TN. It can be seen from this, under the different tree species richness, the value of soil active organic carbon largely depends on the total organic carbon content, which directly participated in the process of soil biochemical conversion and achieved mutual conversion. At the same time, the SOC, MLC, EOC, DOC and MicC also had a significantly positive relationship with root biomass and moisture, but showed extremely significant negative correlation with pH, the reasons as follows: the fine root not only stores a lot of carbon, but also carbon entering the soil through fine root is one or more times to the input of above-ground litter. Under properly conditions, it can increase the storage of soil organic carbon by decomposition effects of soil microbes; to some extent, high moisture can increase the activity of microorganism, however, low value of pH can inhibit their activity, which is harmful for the turnover of nutrients in the soil and thus decrease the value of soil organic carbon storage.

5 CONCLUSIONS

The results revealed that the concentrations of SOC were higher in the mixed combinations than that in monoculture, but mainly on the surface layer (0-15 cm). The MLC, EOC and DOC fluctuated along the gradient of species richness in this natural forest. However, the MicC concentration decreased with increasing neighborhood species diversity, which is surprising as the diverse litterfall and fine roots were generally believed to increase the substrate quality, notwithstanding this is in accordance with the decreased annual aboveground litterfall with increasing species diversity. What is more, the ratio of MLC to SOC and ratio of MicC to SOC decreased with neighborhood species diversity, suggesting the increased tree species diversity may increase the SOC stability and thus enhance the soil carbon sequestration potential. Compared to the aboveground litterfall, the fine root biomass showed significant relationships with soil organic carbon and labile carbon fractions.

REFERENCES

Alexander, M. 1961, Introduction to soil microbiology. *Soil Science* 125(5): 447.

Blair, G.J., Lefroy, RDB, Lisle L. 1995. Soil carbon fractions based on their degree of oxidation, and the development of a carbon management index for agricultural systems. *Aust J Agric Res.* 46:1459–1466.

Chen, C.R., Xu, Z.H. 2008, Analysis and behavior of soluble organic nitrogen in forest soils. *Journal of Soils and Sediments.* 2008. 8(6):363–378.

Chen, L.L., Wen, Q.X. 1998. Effect of land use pattern on mineralization of residual C and N from plant materials decomposing under field conditions. *Pedosphere* 8:311–316.

Ge, L.L., Ma, L.H., Bian, J.L., et al. 2013. Effects of returning maize straw to field and site specific nitrogen management on grain yield and quality in rice. *China. Rice Sci.* 27:153–160.

Hassink, J. 1993. Relationship Between the Amount and the Activity of the Microbial Biomass in Dutch Grassland Soils: Composition of the Fumigation-incubation Method and the Substrate-induced Respiration Method. Soil Biol. *Biochem.* 25:530–538.

Haynes, R.J. 2005. Labile organic matter fractions as central components of the quality of agricultural soils: an overview. *Adv. Agron.* 85: 221–268.

Jenkinson, D.S.1981. Microbial biomass in soil: Measurement and turnover. *Soil Biochemistry* 5: 415–472.

Jiang, P.K., Xu, Q.F., Xu, Z.H.& Cao, Z.H. 2006. Seasonal changes in soil labile organic carbon pools within a Phyllostachys praecox stand under high rate fertilization and winter mulch in subtropical China. *For. Ecol. Manage.* 236:30–36.

Kandeler, E., Tscherko, D., Spiegel, H. 1999. Long-term monitoring of microbial biomass, N mineralization and enzyme activities of a chernozem under different tillage management. *Biol. Fertil. Soils.* 28:343–351.

Lal, R. 2004. Soil carbon sequestration impacts on global climate change and food security. *Science.* 304:1623–1627.

Loveland, P., Webb, J. 2003. Is there a critical level of organic matter in the agricultural soils of temperate regions: a review. *Soil Tillage Res.* 70(1):1–18.

Moebius-Clune B DJ, Gugino BK, et al. Comprehensive assessment of soil health: The Cornell Framework Manual, Edition 3.1, 2016, Cornell Univ., Ithaca, NY.

Motavalli PP, Palm CA, Parton WJ, et al. Comparison of laboratory and modeling simulation methods for estimating carbon pools in tropical forest soil. *Soil Biology and Biochemistry*, 1994, 26(8): 935–944.

Scaglia B, Adani F. Biodegradability of soil water soluble organic carbon extracted from seven different soils. *Journal of Environmental Sciences*, 2009, 21(5): 641–646.

Schlesinger WH. 1982. Carbon storage in the caliche of arid soils: A case study from Arizona. *Science* 133: 247–255.

Shen H, Cao ZH. Effect of fertilization on soil available carbon and carbon availability under different agroecosystems. *Trop. Subtrop. Soi*ls, 1998, 7:1–5.

Shen H, Cao ZH. Study of carbon pool management index in soils under different agroecosystems. *Acta Natural Resourc.*1999, 14:206–211.

Sparling GP. Ratio of Microbial Biomass C to Soil Organic C as a Sensitive Indicator of Changes in Soil Organic Matter. *Aust. J. Soil Res.* 1992, 30:195–207.

Spedding TA, Hamel C, Mehuys GR, Madramootoo CA. Soil microbial dynamics in maize-growing soil under different tillage and residue management systems. Soil Biol. *Biochem.* 2004, 36: 499–512.

Wade J, Culman S W, Hurisso TT, et al. Sources of variability that compromise mineralizable carbon as a soil health indicator: *Soil Science Society of America Journal*, 2018, 82: 243–252.

Weil RR, Islam KR, Stine MA, Gruver JB, Samson-Liebig SE. 2003. Estimating active carbon for soil quality assessment: A simplified method for laboratory and field use. *Am J, Altenat Agric* 18:3–17.

The Role of Agriculture in Climate Change Mitigation – Pawłowski, Litwińczuk & Zhou (eds)

Stability of particulate organic matter from the coastal soils in North China: Indication of moieties and stable isotopic ratios

H. Zhang, Y. Fei & J. Wang
Key Laboratory of Soil Contamination Bioremediation of Zhejiang Province, School of Environmental and Resource Sciences, Zhejiang A & F University, Hangzhou, China

X. Liu
Key Laboratory of Coastal Environment Processes and Ecological Remediation, Yantai Institute of Coastal Zone Research, Chinese Academy of Sciences, Yantai, China

Y. Luo
Key Laboratory of Coastal Environment Processes and Ecological Remediation, Yantai Institute of Coastal Zone Research, Chinese Academy of Sciences, Yantai, China
Nanjing Institute of Soil Science, Chinese Academy of Sciences, Nanjing, China

1 INTRODUCTION

Soil plays an important role in territorial carbon cycling (Bartniczak et al. 2019, Richard & Alfred 2016, Kostecka et al. 2018, Shaw 2019). Understanding the role of organic matter in maintaining a healthy soil is essential for developing ecologically sound agricultural practices. Particulate organic matter (POM) is a type of biologically and chemically active fraction of soil organic matter which was defined operationally as organic particles 0.053 to 2 mm in size (Cambardella & Elliott 1992). The POM was consisted of free particulate organic matter (fPOM) which covered primarily on the surface of soil aggregate, while the intra-aggregate particulate organic matter (iPOM) which was sequestrated inside of soil aggregate. Both of the fPOM and iPOM fractions were poor stability and had a rapid turnover in comparison with the mineral associated organic matter (mSOM) which was physically and chemically protected by soil clay and silt. However, the stability of the different POM also varied with their sources, which could be identified by carbon stable isotopic ratio and the structural functional groups. The objectives of this study were to characterize the fractions of particulate organic matter sourced from different land using in the coast of Yellow River Delta region in North China and to identify the stability of the POM.

2 MATERIALS AND METHODS

Soil samples were collected from three typical land use types including barren area, saltmarsh and farmland in the study area, all the sample information was shown in Table 1. The fractionation of the POM was based the scheme established by Six et al. (2002). The fractions of fPOM and iPOM with different particle size were collected and the mSOM fraction was also collected as a comparison. Solid state nuclear magnetic resonance (SSNMR) and carbon stable isotopes were analyzed for the collected POM fractions and mSOM fraction.

3 RESULTS AND DISCUSSION

The analysis of total soil carbon content for the samples indicated that it presented an increasing trend from barren area to saltmarsh and decreased in farmland upland soils (Table 1). Figure 1 shows the difference of carbon moieties among the fPOM sourced from different land using types by using 13C SSNMR. It turned out two groups based on the characteristics

Table 1. The testing soil samples for POM fractionation.

Sample No.	Land type	Vegetation or crops	Location	SOC (g/kg)
BS1	Barren area	Vegetation invisible	Lijing county, Dongying	3.3
BS2	Barren area	Vegetation invisible	Lijing county, Dongying	2.9
BS2	Barren area	Oil field	Gudong district, Dongying	6.4
WS1	Saltmarsh	Suaeda heteropter Kitag	Lijing county, Dongying	2.9
WS2	Saltmarsh	Tamarix chinensis Lour	Hekou district, Dongying	4.5
WS3	Saltmarsh	Phragmites communis Trin	Hekou district, Dongying	45.2
WS4	Saltmarsh	Spartina alterniflora	Gudong district, Dongying	21.6
FS1	Farmland	Cotton	Lijing county, Dongying	6.9
FS2	Farmland	Wheat	Zhanhua county, Binzhou	19.7
FS3	Farmland	Corn	Wuli county, Binzhou	26.4

Note: SOC indicates soil organic carbon

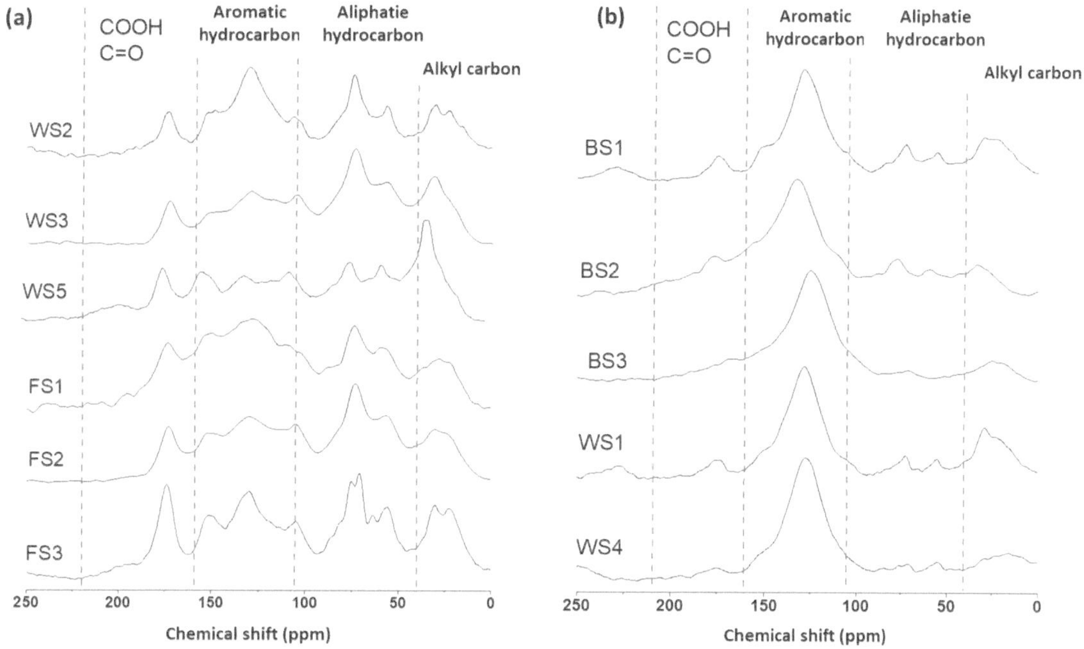

Figure 1. The 13C SSNMR spectra indicated that carbon moieties of the POM varied with different land using, which might be affected by their carbon sources.

of the NMR spectra. The first group including fPOM from farmlands and the wetland of Tamarix chinensis Lour and Phragmites communis Trin, which indicates a prominent peaks of aliphatie hydrocarbon in the moieties however lower intensity of aromatic hydrocarbon moieties. The other group comprised of the fPOM from barren area and the wetland of Suaeda heteropter Kitag and Spartina alterniflora, which shows a prominent peak of aromatic hydrocarbon moieties.

The δ13C values of different soil organic matter fractions mainly followed the order of fPOM< iPOM < mSOM, which implied that free POM (e.g. lignin) has a higher bioavailability than the clay protected organic carbon regarding the decomposition by microorganisms (Six. et al. 2000). A significant difference of δ13C values also shown among the different land using. The lowest δ13C value occurred in the wetlands while the highest occurred in barren

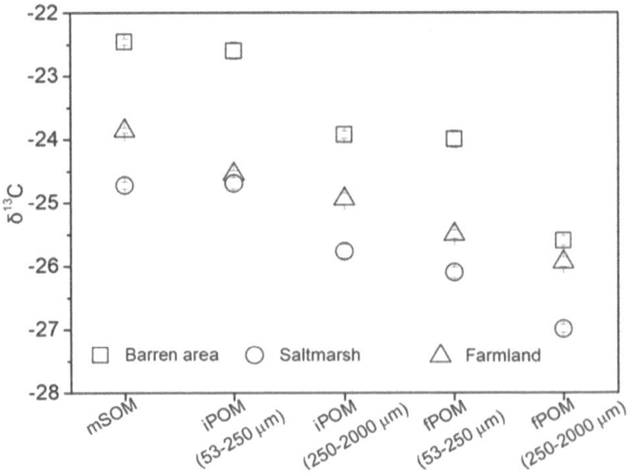

Figure 2. The change of stable carbon isotopic (δ13C) values among the collected POM and mSOM from the soil of different land using types.

area. The trend of δ13C value changed with land using probably was affected by the carbon sources. The δ13C value of marine sourced organic carbon was usually higher than that of terrestrial sourced organic carbon (Cai 1994, Lamb et al. 2006). Therefore the high δ13C value of the POM (or mSOM) in the barren land was mainly affected by the input of marine sourced organic matter carried by tide. The soil organic matter in the saltmarsh has the lowest δ13C value as shown in Figure 2, this could be resulted from the low decomposition under the waterlogging condition in saltmarsh (Cao & Cel 2015, Kujawska & Cel, 2016). While the decomposition could be accelerated after the saltmarsh transferred to farmland due to the cultivation (Hou et al. 2011).

In summary, the results implied that although the content of soil organic carbon was relatively lower in barren area, it has a higher stability in the soils. In the contrast, the contents of soil organic carbon in the soils of saltmarsh were high, however a relative high proportion of labile fractions in the organic matter indicated their instability.

The soil organic matter in the corn and wheat lands had a higher stability due to the higher proportion of stable organic matter fractions due to the long-term cultivation. All these results demonstrated that the properties of the organic matter is as important as the content in the different soils. Meanwhile, soil organic matter of the saltmarsh should be of concerned since it may have a higher tendency to releasing greenhouse gases to the atmosphere than the soils of the dryland in North China.

Therefore, studies on the properties rather than content singly of the organic matter in the different soils should be of concerned greatly in the future.

REFERENCES

Bartniczak, B. & Raszkowski A. 2019. Sustainable development in Asian countries - Indicator-based approach. *Problems of Sustainable Development/Problemy Ekorozwoju.* 14(1):29–42.

Benoit, P., Madrigal, I., Preston, C., et. 2008. Sorption and desorption of non-ionic herbicides onto particulate organic matter from surface soils under different land uses, *European Journal of Soil Science* 59:178–189.

Besnard, E., Chenu, C., Balesdent J., et al. 1996. Fate of particulate organic matter in soil aggregates during cultivation, *European Journal of Soil Science* 47:495–503.

Cambardella, C., Elliott E. 1992. Particulate soil organic-matter changes across a grassland cultivation sequence, *Soil Science Society of America Journal* 56:777–783.

Cao, Y.C. & Cel, W. 2015. Sustainable Mitigation of Methane Emission by Natural Processes. *Problems of Sustainable Development/Problemy Ekorozwoju* 10(1):117–121.

Guo, X., Luo, L., Ma, Y., et al. 2010. Sorption of polycyclic aromatic hydrocarbons on particulate organic matters, *Journal of Hazardous Materials* 173:130–136.

Hou, C. C., Song, C. C., Li, Y. C., et al. 2011. Seasonal dynamics of soil organic carbon and active organic carbon fractions in *Calamagrostis angustifolia* wetlands topsoil under different water conditions (in Chinese). *Environmental Science* 32(1):290–297.

Kostecka, J., Garczyńska, M. & Pączka, G. 2018. Food waste in the organic recycling system and sustainable development. *Problems of Sustainable Development/Problemy Ekorozwoju* 13(2):157–164.

Kujawska, J. & Cel, W. 2016. Mitigation of greenhouse effect by reduction of the methane emissions. *Problems of Sustainable Development/Problemy Ekorozwoju* 11(2):127–129.

Lindzen R., Sloan A.P. 2016. Global Warming and the Irrelevance of Science. *Problems of Sustainable Development/Problemy Ekorozwoju* 11(2):119–125.

Shaw, K. 2019, Implementing sustainability in global supply chain. *Problems of Sustainable Development/ Problemy Ekorozowju* 14(2):117–127.

Six, J., Paustian, K. Elliott, E. T. et al. 2000. Soil structure and organic matter. I. Distribution of aggregate-size classes and aggregate-associated carbon. *Soil Science Society of America Journal* 64(2):681–689.

The Role of Agriculture in Climate Change Mitigation – Pawłowski, Litwińczuk & Zhou (eds)

Stand structure and mass-ratio effect more than complementarity effect determine productivity in subtropical forests

S. Ouyang, W. Xiang & P. Lei
Faculty of Life Science and Technology, Central South University of Forestry and Technology, Changsha, Hunan, China
Huitong National Station for Scientific Observation and Research of Chinese Fir Plantation Ecosystems in Hunan Province, Huitong, Hunan, China

1 INTRODUCTION

Forests play an important role in global carbon (C) sequestration through the absorption of 35% of CO_2 emissions from fossil-fuel and account for nearly 25% of C sinks of global terrestrial ecosystems (King et al. 2012). Hence, a better understanding of the relationship between biodiversity and ecosystem functioning (BEF) is critical to sustain forest ecosystem functions and services, in particular regarding aspects such as C storage, productivity, nutrient cycling or ecosystem stability (Liang et al. 2016, Liu et al. 2018). The spatial distribution of forest productivity across landscapes has been well reported (e.g. Guo & Ren 2014), however, the underlying mechanisms of driving productivity are still not well understood in forest ecosystems at landscape scales.

Over the last decade, there has been increasing concern about the effects of biodiversity loss on ecosystem functioning (Liang et al. 2016, Liu et al. 2018, Bartniczak & Raszkowski 2019). At present, two hypotheses were proposed to explain the effects of biodiversity on forest productivity: the complementarity effect and sampling effect. The complementarity effect assumes that increasing diversity enhances forest productivity through niche differentiation (increased resource-use efficiency) (Tilman et al. 1997). Thus, if complementarity effect is indeed the underlying driver, a positive correlation should be found between ecosystem functioning and functional diversity (FD) (Zhang et al. 2012, Fotis et al. 2017, Yu et al. 2019, Xu 2018). In contrast, the sampling effect proposes that higher species richness (SR) increases community productivity through an increased chance of possessing highly productive species (Tilman et al. 1997, Hooper et al., 2005). The mass-ratio effect is used as a proxy of the selection effect and can be tested by determining whether variation in ecosystem functioning (e.g. forest productivity in this study) is related to the community-weighted mean (CWM) of species traits (Chiang et al., 2016, Fotis et al. 2017). For example, strong positive relationships were found between functional trait values between maximum tree height and biomass (Chiang et al. 2016, Fotis et al. 2017). These two mechanisms are not exclusive and may contribute to BEF relationships simultaneously (Hooper et al. 2005, Wu et al. 2015, Chiang et al. 2016), though the relative importance of complementarity and the sampling effect is still controversial (Chiang et al. 2016, Fotis et al. 2017).

The majority of BEF studies of forest systems have reported positive relationships between SR and forest productivity at regional and continental scales, although negative and non-significant BEF relationships have also been reported (Forrester & Bauhus 2016, Liang et al. 2016). Based on more than 777, 000 permanent plots in 44 countries, the results showed a general positive correlation between biodiversity and stand productivity (Liang et al. 2016). Most studies to date were conducted in relatively species-poor forest ecosystems (e.g. temperate and boreal forests) (e.g. Paquette & Messier 2011, Fotis et al. 2017), with fewer studies in the subtropics (e.g. Barrufol et al. 2013, Wu et al. 2015).

The widely accepted positive BEF has been found in forests across the world, but other factors (i.e. climatic, edaphic, stand structure and topographic factors) could have a much larger impact on biomass or productivity (Zhang & Chen 2015, Forrester & Bauhus 2016, Ouyang et al. 2016). Studies in natural forests reveal that the strength of BEF can be strongly confounded by climate, site conditions and stand structure (Zhang & Chen 2015, Forrester & Bauhus 2016). For example, forests in poor soil quality sites have been showed to exhibit stronger positive diversity effects on productivity than stands on high soil fertility sites (Toïgo et al. 2015). Stand age is also an important driver for ecosystem functions (e.g. biomass and productivity) but is not often included in BEF studies of forests (Forrester & Bauhus 2016, Liu et al. 2018). To date, few BEF studies have included all these factors simultaneously (as opposed to one or two confounding factors) for a systematic examination of their confounding effects. Therefore, it is crucial to consider these factors when testing multivariate and indirect relationships between diversity and productivity in forest ecosystems.

Subtropical forests are abundant with rich tree species and characterized by complex stand structure and various environmental conditions (Xiang et al. 2016). However, whether and to what degree biodiversity and other factors (climate, soil quality and stand characteristics) influence productivity in diverse subtropical forests is not well understood. In this study, we analyzed a large dataset from permanent forest inventory plots across Hunan Province to examine the potential factors (i.e. forest structure and environmental variables) that may affect the relationship between diversity and forest productivity. We addressed the following questions: (1) Do complementarity or mass ratio effects contribute more to forest productivity in subtropical forests? and (2) What is the relative importance of environmental factors, stand structure and diversity in determining stand productivity of subtropical forests?

2 METHODS

2.1 *Study area*

The study was conducted in Hunan Province (latitude 108° 47′ E-114° 15′ E, longitude 24° 38′ N-30° 08′N) situated in the mid-subtropical area of China (Figure 1). Hunan Province is located at the transition zone from the Yunnan-Guizhou plateau to the lower mountains and hills on the southern side of the Yangtze River, has an elevation

Figure 1. Hierarchical partitioning of the variation explained for forest productivity by each predictor. These predictors are forest structure [maximum DBH (maxDBH), stand age and stand density], diversity (richness and functional diversity), CWM for maximum tree height (CWM Hmax) and site quality index.

of 21-2122 m above sea level. A typical humid mid-subtropical monsoon covers this region with average annual air temperature is 17°C. The annual precipitation is about of 1200-1700 mm, occurring primarily between April and October. The average annual air temperature is 16-18°C (Huang et al. 2013). The soils are red-yellow podzolic soils, which developed mostly from slate and shale parent material, and are classified as Plinthudults, according to the US Soil Taxonomy.

2.2 *Data collection and productivity calculations*

Forest stand data came from the permanent sample plots used for the National Forest Inventory of China (CNFI) in Hunan Province, which were surveyed in both the sixth and seventh CNFI (6CNFI, 2009, and 7CNFI, 2014, respectively). For each forest plot, geographic location (latitude, longitude), topographic variables (altitude, slope, aspect, slope position) and soil variables (depth, texture, type and gravel content) were collected. Species name and each individual with the diameter at breast height (DBH) ≥ 5cm were documented. The forest productivity of each plot was determined as biomass increments from 2009 to 2014. Total biomass of each plot was estimated using the biomass expansion factor method (Fang et al. 2014). For more detailed information on the productivity calculations, see Ouyang et al. (2019).

2.3 *Functional traits and biodiversity indices*

We calculated three aspects of biodiversity: species diversity and functional diversity (FD) to compare their ability to explain productivity. Species diversity was quantified as species richness (SR, species number in a given plot). FD describes the variability of characteristics of the species in a community. For more detailed information on functional traits sources and the FD, see Ouyang et al. (2019).

2.4 *Stand structural variables*

Stand structural variables used in this study included mean DBH, maximum DBH and stand density (Chiang et al. 2016, Forrester & Bauhus 2016) and stand age (Zhang & Chen 2015).

2.5 *Site quality index*

The site quality index of each site, an index of the nutrient supply, was determined based on several factors including topographical factors (altitude, slope, aspect, and slope position) and soil factors (texture, depth, gravel content, bedrocks, bulk density) by cluster analysis and discriminant analysis (Courtin et al. 1988, Zhang & Chen 2015).

2.6 *Statistical analysis*

The normality of the variables was tested, and then the variables were ln-transformed and standardized prior to all analyse. We used step-wise multiple regression analysis (lm method in R) to simultaneously evaluate the effects of all predictors based on the Akaike information criterion (AIC) (Quinn and Keough, 2002) using the R package 'MASS'. The first full model included site index, four forest structure variables (mean DBH, maximum DBH, stand density and stand age), two biodiversity indices (SR and FD) and two CWM traits for maximum tree height (CWM Hmax) and wood density (CWM WD). We then used hierarchical partitioning to quantify the percentage of variation explained by the different predictor variables when they were all included together in the final multiple regression, using R package 'hier.part' (Walsh and Nally, 2013).

After the main factors were determined above, structural equation modeling (SEM) was used to assess the direct and indirect effects of the main driving factors on forest productivity. We parameterized our SEM model and tested its goodness of fit using the goodness-of-fit index (GFI), Chi-square test and root mean square error of approximation (RMSEA). Requirements for a good model included an insignificant Chi-square test statistic with a P value > 0.05 (indicating that expected and observed covariance matrices are not statistically different), GFI values and RMSEA > 0.95 and < 0.08, respectively (Zhang & Chen et al., 2015). Furthermore, we determined the total effects (direct plus indirect effects) of the main driving factors on forest productivity. The SEM model was performed using the R package 'lavaan' (Rosseel, 2012). All the statistical analyses were implemented in R version 3.4.1 (R Development Core Team 2016).

3 RESULTS

3.1 *Hierarchical partitioning of variation analysis: The main drivers of forest productivity*

The final multiple regression model based on AIC showed that environmental factor (soil quality index), forest structural factors (maximum DBH, stand age and stand density), CWM Hmax and diversity (SR and FD) exhibited a significant influence on forest productivity (Table 1).

The hierarchical partitioning of variation analysis showed the relative importance of each driving factor in decreasing order: stand density (25.86%) > stand age (21.35%) > CWM Hmax (19.46%) > SR (11.97%) > maximum DBH (9.12%) > FD (6.52%) > SQI (5.72%) (Figure 1). We found that forest structural factors (stand age and stand density) were more important than other driving factors for forest productivity.

3.2 *SEM model: The direct and indirect effects of main drivers on forest productivity*

The SEM model accounted for 46% of the variation in forest productivity (Figure 2). SQI, stand age, stand density, FD and CWM Hmax had significant direct positive impacts on forest productivity, in which the standardized direct effects were 0.13, 0.29, 0.34, 0.25 and 0.29, respectively. Meanwhile, SQI, stand age and stand density also presented indirect effects on forest productivity.

The standardized total effects of each driving factor decreased in the order: stand density (0.40) > stand age (0.35) > CWM Hmax (0.29) > FD (0.25) > SQI (0.21) (Figure 3). The results of standardized total effects also indicated that stand density and stand age were the most important driving factors of forest productivity.

Table 1. Multiple linear models for predicting forest productivity (ln-transformed) from environment factors: site index, forest structural factors (maximum DBH, stand age and stand density), and diversity factors (species richness and functional diversity), CWM Hmax as the community-weighted mean (CWM) trait value for tree maximum height. Site index included three categories: poor, medium, good). Explanatory variables were excluded based on AIC. SS is the sums of square.

Variables	df	Type III SS	F-value	P-value
Site index	2	6.52	39.86	0.000
Maximum DBH	1	6.01	73.41	0.000
Stand age	1	9.75	119.1	0.000
Stand density	1	18.18	222.15	0.000
CWM Hmax	1	4.99	60.93	0.000
Species richness	1	2.51	30.65	0.000
Functional diversity	1	1.43	17.42	0.000

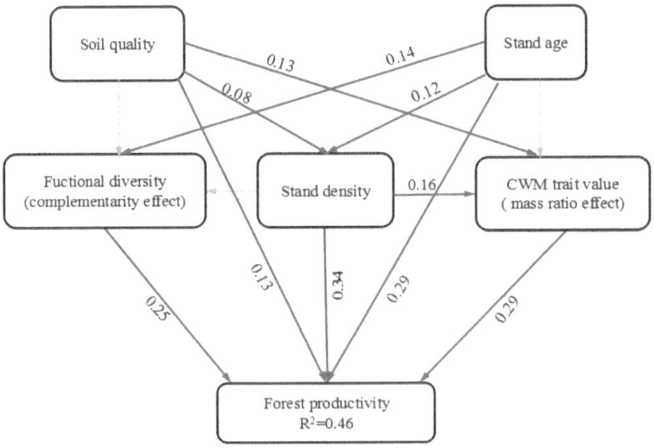

Figure 2. Structural equation models linking the relative importance of site quality index, structural factors (stand age and stand density), functional diversity (niche complementarity effect) and CWM trait value for maximum height (mass ratio effect) on forest productivity. The coefficients are standardized prediction coefficients for each causal path. Solid lines represent significant paths ($p \leq 0.05$) and dash lines for non-significant paths ($p > 0.05$).

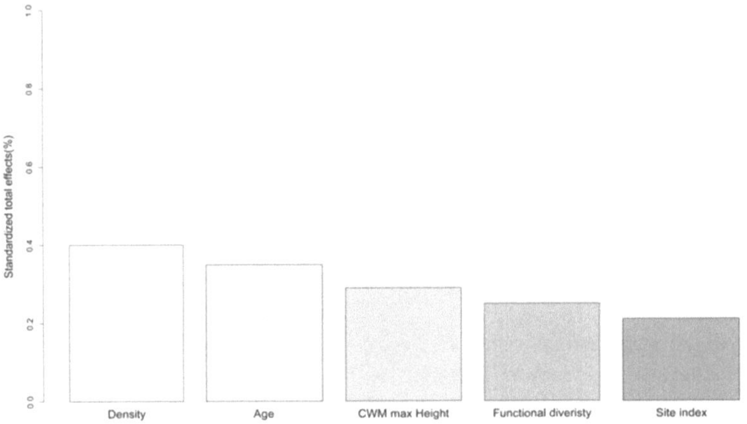

Figure 3. Standardized total effects (sum of direct and indirect effects) derived from the SEM model.

4 DISSCUSSION

4.1 *Positive effects of biodiversity on productivity and underlying mechanisms in subtropical forests*

Positive effects of diversity on forest productivity have been reported by recent broad-scale studies (Zhang et al. 2012, Liang et al. 2016), and most of them were conducted in temperate or boreal forests. The relationship between biodiversity and forest productivity has received less attention in subtropical forests, especially at a large scale. A significantly positive (Liu et al. 2018) and no-significant (Wu et al. 2015) effect of SR on biomass (carbon storage) were observed in subtropical forest. These different relationships between SR and biomass can vary depending on the scales used. Recent studies found that both productivity and biomass were usually positively related to SR at a small spatial scale (0.04 ha), but became neutral or negative relationships at larger scales (0.25 or 1 ha) from temperate to tropical forests

(Chisholm et al. 2013, Wu et al. 2015). The significantly positive relationships between different aspects of diversity (i.e. SR and FD) and forest productivity that we are in according with these findings given the similar scale of the analysis in this study (plots with a scale of 0.07 ha). Chisholm et al. (2013) ascribed the neutral or negative relationships between SR and biomass at larger scales to the strong effects of environmental factors (e.g. soil fertility). We included site index in our models, and therefore partly considered for the effects of habitat heterogeneity (e.g. light, soil moisture and nutrient availability). Nonetheless, we still found the significantly positive effects of diversity on forest productivity, even after the effects of covering variables were accounted for.

An independent relationship between FD and forest productivity indicates that niche complementarity is an underlying mechanism promoting productivity in our subtropical forests. FD was also positively correlated with forest productivity after accounting for other predictor variables. This result further supports the notion that niche complementarity plays a significant role in determining the variation in productivity in this subtropical forest. These results are also in line with those of Paquette and Messier (2011), who found that FD did significantly affect the productivity in boreal forests. Chiang et al. (2016) found that while FD was negatively associated with biomass and productivity in subtropical forests, FD still had an indirect positive effect on biomass and productivity by influencing stem density, suggesting that niche complementarity still contributes to ecosystem functions in these forests. The mass ratio hypothesis predicts that ecosystem properties should be largely determined by the characteristics of the dominant species within a community (Grime 1998). As a consequence, we expected to see a significant effect of the community-weighted mean (CWM) of species traits on ecosystem functions if the mass ratio effect was supported (Hooper et al. 2005, Finegan et al. 2015, Chiang et al. 2016). We found that CWM trait value for tree maximum height was directly related to forest productivity. Similar results were reported by Chiang et al. (2016) who noted that CWM trait value for tree maximum height was the best predictor of wood productivity in a subtropical forest. Moreover, Finegan et al. (2015) found strong MR effects on aboveground biomass storage and increment in tropical forests located in Bolivian, Costa Rican, and Brazilian. Our results support the notion that mass ratio and complementarity effects are not mutually exclusive in these forests (Hooper et al. 2005, Wu et al. 2015). Furthermore, we found the mass ratio effect was stronger than the complementarity effect in our study. Our results provide additional evidence for a positive diversity-productivity relationship at a relatively small scale (0.07 ha), where community processes (e.g. sampling effect) play a dominant role at this scale (Chisholm et al. 2013). Mass ratio effects are likely observed to dominate in fertile regions (i.e. temperate forest), and functional diversity (complementarity effect) may play a larger role in determining ecosystem functions in harsher environments (i.e. boreal forests) (Paquette & Messier et al. 2011).

4.2 *The effects of environmental factors and stand structure on forest productivity*

Many studies reveal that the BEF relationships vary markedly depending upon environmental conditions (i.e. edaphic and topographic conditions) and stand structure (e.g. DBH, stand density and stand age) in natural forests, showing positive, unimodal, negative, or non-significant relationships (Jiang et al. 2009, Jucker et al. 2016, Fotis et al. 2017), indicating that environmental conditions and stand structure may play significant roles in generating variation in forest productivity. However, the relative roles of environmental factors and stand structure effects on forest productivity at a large-scale have received little attention (but see Paquette & Messier 2011, Liang et al. 2016).

Both the hierarchical partitioning of variation analysis and SEM model showed that stand density was the main structural variable modulating the productivity of these subtropical forest ecosystems, followed by forest age. Recent studies have shown that high stand density can have strong effects on forest productivity (Guo & Ren 2014, Forrester & Bauhus 2016, Jucker et al. 2016). At low stand densities, trees are likely to be too far apart for complementarity effects to be significant (Forrester & Bauhus 2016). As stand density increases, complementarity effects have been found to increase (Boyden et al. 2005) or decrease (Boyden et al. 2005). Stand density has also shown to be a stronger determinant of forest productivity than SR based on large inventory data sets (Vilà et al. 2013, Guo & Ren 2014). Our results, together with these studies, indicate the importance of taking stand density into consideration when analyzing the relationship between diversity and productivity.

Forest age was another major modulator of forest productivity in our study, and was also an important predictor for plant carbon turnover times for all forest types in the Zhejiang Province, Southern China (Xu et al. 2018). The strong positive effects of stand age on biomass were attributed to cumulative tree growth over time (Poorter et al. 2016). Stand age may also directly affect biomass and productivity via changing stand structure (e.g. density in our study) and/or species diversity as forests develop (Zhang & Chen, 2015). Many studies have shown that forest productivity increases during early successional stages, and then declines as the forests mature (e.g. Guo & Ren 2014). Notably most of the forests in our study are still young (age < 45 years), indicating these forests also have the potential to sequester great amounts of carbon through forest growth in the future. In according with previous studies (Fichtner et al. 2015, Fotis et al. 2017), our results found that maximum DBH (large-diameter trees) had strong positive associations with forest productivity, even when controlling for other predictor variables in the final multiple regression model. Thus, future studies about the role of diversity on productivity should account for these important stand structural attributes.

5 CONCLUSIONS

Much of the controversy has centered on the BEF relationship over the past two decades (Liang et al. 2016, Fotis et al. 2017). Our findings reveal that in addition to biodiversity, stand structure and environmental factors (e.g. soite index) contribute to the geographic variation in productivity in subtropical forests. We found that both the complementarity effect and mass-ratio effect contributed to the positive effects on forest productivity simultaneously, although the relative importance of the mass-ratio effect was larger than the complementarity effect in our study. Overall, stand structural attributes were the major determinants of forest productivity. Our results suggest that stand structure (e.g. stand density and forest age) is far more important for determining productivity than biodiversity, at least for these subtropical forests, implying the importance of accounting for stand structure when analyzing the effects of biodiversity on forest ecosystem functions. We suggest that when considering the productivity and carbon sequestration ability of subtropical forests, biodiversity should not be considered without also considering stand structure and management.

ACKNOWLEDGEMENTS

This work was supported by the National Key Research and Development Program of China (2016YFD0600202), National Natural Science Foundation of China (31570447 and 31700636), China Postdoctoral Science Foundation (2017M612605), and the Huitong Forest Ecological Station Funds by the State Forestry and Grass Administration of China.

REFERENCES

Barrufol, M., Schmid, B., Bruelheide, H., Chi, X., Hector, A., Ma, K., Michalski, S., Zhiyao, T., Niklaus, P.A. 2013. Biodiversity promotes tree growth during succession in subtropical forest. *Plos One* 8, e81246.

Bartniczak, B. & Raszkowski, A. 2019. Sustainable Development in Asian countries – Indicator-based Approach. *Problemy Ekorozwoju/Problems of Sustainable Development* 14(1):29–42.

Boyden, S., Binkley, D., & Senock, R. 2005. Competition and facilitation between Eucalyptus and nitrogen-fixing Falcataria in relation to soil fertility. *Ecology* 86: 92–1001.

Brophy, C., Dooley, Á., Kirwan, L., Finn, J.A., McDonnell, J., Bell, T., ... Connolly J. 2017. Biodiversity and ecosystem function: Making sense of numerous species interactions in multi-species communities. *Ecology* 98: 1771–1778.

Cardinale, B.J., Duffy, J.E., Gonzalez, A., Hooper, D.U., Perrings, C., Venail, P., ... Naeem, S. 2012. Biodiversity loss and its impact on humanity. *Nature* 486: 59–67.

Chiang, J.M., Spasojevic, M.J., Muller-Landau, H.C., Sun, I.F., Lin, Y., Su, S.H., ... McEwan, R.W. 2016. Functional composition drives ecosystem function through multiple mechanisms in a broadleaved subtropical forest. *Oecologia* 182: 829–840.

Chisholm, R. A., & Zimmerman J K. 2013. Scale - dependent relationships between tree species richness and ecosystem function in forests. *Journal of Ecology* 101: 1214–1224.

Courtin, P.J., Klinka, K., Feller, M.C., & Demaerschalk, J.P. 1988. An approach to quantitative classification of nutrient regimes of forest soils. *Canadian Journal of Botany-Revue Canadienne De Botanique* 66: 2640–2653.

Faith, D.P. 1992. Conservation evaluation and phylogenetic diversity. *Conservation Biology* 61: 1–10.

Fang, J., Guo, Z., Hu, H., Kato, T., Muraoka, H., & Son, Y. 2014. Forest biomass carbon sinks in east Asia, with special reference to the relative contributions of forest expansion and forest growth. *Global Change Biology* 20: 2019–2030.

Fichtner, A., Forrester, D. I., Härdtle, W., Sturm, K., & von Oheimb, G. 2015. Facilitative-competitive interactions in an old-growth forest: The importance of large-diameter trees as benefactors and stimulators for forest community assembly. *Plos One* 10, e0120335.

Finegan, B., Peña-Claros, M., de Oliveira, A., Ascarrunz, N., Bret-Harte, M.S., Carreno-Rocabado, G., ... Poorter, L. 2015. Does functional trait diversity predict above-ground biomass and productivity of tropical forests? Testing three alternative hypotheses. *Journal of Ecology* 103:191–201.

Fotis, A.T., Murphy, S.J., Ricart, R.D., Krishnadas, M., Whitacre, J., Wenzel, J.W., & Comita, L.S. 2017. Aboveground biomass is driven by mass-ratio effects and stand structural attributes in a temperate deciduous forest. *Journal of Ecology* 106: 561–570.

Forrester, D.I., & Bauhus, J. 2016. A review of processes behind diversity–productivity relationships in forests. *Current Forestry Reports* 2: 45–61.

Grime, J.P. 1998. Benefits of plant diversity to ecosystems: immediate, filter and founder effects. *Journal of Ecology* 86: 902–910.

Guo, Q. & Ren, H. 2014. Productivity as related to diversity and age in planted versus natural forests. *Global Ecology and Biogeography* 23:1461–1471.

Hooper, D.U., Chapin, F.S., Ewel, J.J., Hector, A., Inchausti, P., Lavorel, S., ... Wardle, D.A. 2005. Effects of biodiversity on ecosystem functioning: a consensus of current knowledge. *Ecological Monographs*, 75: 3–35.

Huang, J., Sun, S., Xue, Y., & Zhang, J. 2013. Spatial and temporal variability of precipitation indices during 1961–2010 in Hunan Province, central south China. *Theoretical and Applied Climatology* 118: 581–595.

Jucker, T., Avăcăriței, D., Bărnoaiea, I., Duduman, G., Bouriaud, O., & Coomes, D.A. 2016. Climate modulates the effects of tree diversity on forest productivity. *Journal of Ecology* 104: 388–398.

King A.W., Hayes D.J., Huntzinger D.N., Huntzinger, D.N., West, T.O., & Post, W.M. 2012. North American carbon dioxide sources and sinks: magnitude, attribution, and uncertainty. *Frontiers in Ecology and the Environment* 10:512–519.

Lasky, J.R., Uriarte, M., Boukili, V.K., Erickson, D.L., John Kress, W., & Chazdon, R.L. 2014. The relationship between tree biodiversity and biomass dynamics changes with tropical forest succession. *Ecology Letters* 17:1158–1167.

Liang, J., Crowther, T.W., Picard, N., Wiser, S., Zhou M., Alberti, G., ... Reich, P.B. 2016. Positive biodiversity-productivity relationship predominant in global forests. *Science* 354: 6309.

Liu X., Trogisch S., He J., Niklaus, P.A., Bruelheide H., Tang Z., ... Ma K. 2018. Tree species richness increases ecosystem carbon storage in subtropical forests. *Proceedings of the Royal Society B* 285: 20181240.

Ouyang, S., Xiang, W., Wang, X., Zeng, Y., Lei, P., Deng, X., & Peng, C. 2016. Significant effects of biodiversity on forest biomass during the succession of subtropical forest in south China. *Forest Ecology and Management* 372:291–302.

Ouyang, S., Xiang, W., Wang, X., Xiao, W., Chen, L., Li, S., ... & Peng, C. (2019). Effects of stand age, richness and density on productivity in subtropical forests in China. *Journal of Ecology* 107:2266–2277.

Paquette, A. & Messier, C. 2011. The effect of biodiversity on tree productivity: from temperate to boreal forests. *Global Ecology Biogeography* 20:170–180.

Poorter, L., Bongers, F., Aide, T.M., Zambrano, A., Balvanera, P., Becknell, J.M. ... Rozendaal, D.A. (2016). Biomass resilience of Neotropical secondary forests. *Nature* 530:211–214.

Potter, K.M. & Woodall, C.W. 2014. Does biodiversity make a difference? Relationships between species richness, evolutionary diversity, and aboveground live tree biomass across US forests. *Forest Ecology and Management* 321:117–129.

Quinn, G.P., & Keough, M.J. 2002. Experimental design and data analysis for biologists. Cambridge University Press, UK.

R Development Core Team. 2016. R: A Language and Environment for Statistical Computing. R Foundation for Statistical Computing, Vienna, Austria.

Rosseel, Y. 2012. lavaan: an R package for structural equation modeling. *Journal of Statistical Software*, 48:1–36.

Ruiz-Jaen, M.C., & Potvin, C. 2011. Can we predict carbon stocks in tropical ecosystems from tree diversity? Comparing species and functional diversity in a plantation and a natural forest. *New Phytologist*, 189:978–987.

Tilman, D., Knops, J., Wedin, D., Reich, P., Ritchie, M., & Siemann, E. 1997. The influence of functional diversity and composition on ecosystem processes. *Science* 277:1300–1302.

Vilà, M., Carillo-Gavilán, A., Vayreda, J., Bugmann, H., Fridman, J., Grodzki, W., ... Trasobares, A. (2013). Disentangling biodiversity and climatic determinants of wood production. *Plos One* 8, e53530.

Walsh C., & Nally RM. 2013. hier.part: Hierarchical partitioning. See https://cran.r-project.org/web/packages/hier.part/index.html.

Wu, X., Wang, X., Tang, Z., Shen, Z., Zheng, C., Xia, X., & Fang, J. 2015. The relationship between species richness and biomass changes from boreal to subtropical forests in China. *Ecography* 37:602–613.

Xiang, W.H., Zhou, J., Ouyang, S., Zhang. S.L., Lei, P.F., Li, J.X., Deng, X.W., Fang, X., Forrester, D.I. 2016. Species-specific and general allometric equations for estimating tree biomass components of subtropical forests in southern China. *European Journal of Forest Research* 135:963–979.

Xu, J. and Li, J. 2018. The Tradeoff between Growth and Environment: Evidence from China and the United States. *Problemy Ekorozowju/Problems of Sustainable Development* 13(1):15–20.

Yang, Y., Huang, S., Huang, X. 2019. An Empirical Comparison of Environmental Behaviors in China's Public and Private Sectors. *Problemy Ekorozwoju/Problems of Sustainable Development/*, 14(2):101–110.

Zhang, Y., & Chen, H. 2015. Individual size inequality links forest diversity and above-ground biomass. *Journal of Ecology* 103:1245–1252.

Zhang, Y., Chen, H., & Reich, P.B. 2012. Forest productivity increases with evenness, species richness and trait variation: a global meta-analysis. *Journal of Ecology* 100:742–749.

The Role of Agriculture in Climate Change Mitigation – Pawłowski, Litwińczuk & Zhou (eds)
 ISBN 978-0-367-43372-7

Anaerobic co-digestion of brewery spent grain and municipal sewage sludge under mesophilic and thermophilic conditions

A. Szaja, A. Montusiewicz, M. Lebiocka & E. Nowakowska
Faculty of Environmental Engineering, Lublin University of Technology, Lublin, Poland

1 INTRODUCTION

Brewery spent grain (BSG) is a main by-product of brewery industry produced globally in large quantities. It finds several applications as food supplement and cattle feed. Due to its chemical composition and significant biogas potential, it may be considered as a source of renewable energy (Mussato et al. 2006). However, mono-digestion of BSG is not efficient enough mainly due to the demand of long hydraulic retention times and low biodegradability of lignin, one of the BSG component (Gonçalves et al. 2015). The application of different thermal, mechanical and chemical pretreatments has recently been proposed to omit this limitation. Unfortunately, the implementation of these techniques could involve additional investment and operational costs (Sezun et al. 2010). Additionally, it might result in the formation of inhibitory intermediates such as phenolic compounds that could adversely affect the anaerobic digestion process (Retfalvi et al. 2013, Sawatdeenarunat et al. 2015, Panjičko et al. 2017, Kainthola et al. 2019).

On the other hand, the high energetic potential of BSG may be profitable when co-digesting this substrate with sewage sludge (SS). The presence of vitamins, mineral salts and amino acids as well as the beneficial C/N ratio and high buffering capacity of BSG could supplement the composition of SS, significantly improving the biogas production and providing good process stability in exploiting digesters. Simultaneously, the BSG unemployed biogas potential could be harnessed without incurring significant costs for the breweries (Sezun et al. 2010).

At majority of wastewater treatment plants (WWTPs) the anaerobic digestion process is carried out under mesophilic (30-40°C), rather than thermophilic (55-65°C) conditions. The former, provides stable biogas production and requires less energy input as compared to the thermophilic temperature. On the other hand, the thermophilic conditions lead to a higher substrate degradation and ensure a significant reduction of pathogens. In addition, it may result in enhanced biogas production. However, it can also lead to some process instability and generate additional costs, and thus the mesophilic temperatures are still recommended for many substrates (Labatut et al. 2014, Moset et al. 2015, Gebreeyessus & Jenicek 2016). The recent studies on the anaerobic digestion of lignocellulosic wastes have indicated that the thermophilic conditions might result in the higher biogas production, but in many cases the process instability could be observed due to VFA accumulation and inhibitory effect of phenolic compounds (Giuliano et al. 2013, Zhang et al. 2014, Yang et al. 2015, Kainthola et al. 2019).

This study examined the effect of temperature on the anaerobic co-digestion of dried BSG and SS in a batch mode. The co-digestion efficiency was evaluated on the basis of both biogas (BP_{21}) and methane (MBP_{21}) potential as well as kinetics parameters, i.e. biogas (GPR) and methane (MPR) production rate.

2 MATERIALS AND METHODS

2.1 *Material characteristics and sample preparation procedure*

In the present study, the sewage sludge was a main substrate. It was obtained from Lublin municipal wastewater treatment plant (WWTP), Poland. This comprised two-source residues,

Table 1. The main characteristic of substrates used in experiments (the average value and standard deviation are given).

Parameter	Unit	SS	BSG
TS	g kg^{-1}	46.25 ± 0.24	223.9 ± 4.3
VS	g kg^{-1}	34.56 ± 0.24	217.2 ± 4.2
pH		6.62 ± 0.01	-

originating from primary and secondary clarifiers, both thickened. Under laboratory conditions, the samples were mixed at the recommended volume ratio of 60:40 (primary: waste sludge), then homogenized and screened through a 3 mm sieve.

The dried brewery spent grain was used as a co-substrate. It was taken from a local brewery Grodzka 15 in Lublin (Poland), which applied barley for beer production. The sample was collected once as a raw, warm material and then transported to the laboratory. There, it was dried at 55°C for two hours in a laboratory dryer and stored in a hermetic container. The composition of both substrates is presented in Table 1.

The laboratory batch reactors were inoculated with digestate from mesophilic anaerobic digesters operated at Lublin WWTP. In order to ensure anaerobic conditions, their gaseous space was rinsed by nitrogen gas for 2 minutes. The inoculum was post-digested to achieve remnant biogas production smaller than 0.01 Ndm^3 d^{-1}. Then the reactors were supplied with the substrate or substrates according to the assumed schedule, and washed out again by nitrogen gas. The timetable included 30 d for inoculum post-digestion and 21 d for measurements.

2.2 *Laboratory installation and operational set-up*

The study was conducted in batch reactors applying the automatic biogas/methane potential test system BioReactor Simulator (Bioprocess Control AB, Sweden). The laboratory installation consisted of six digesters working in parallel, each with a capacity of 2 dm^3. In order to maintain the assumed temperature, the reactors were placed in thermostatic water bath (18 dm^3). The digesters were equipped with slowly rotating stirrer and a gaseous installation. The volume of generated biogas was continuously monitored by means of a wet gas flow-measuring appliance with a multi-flow cell system. This device employed the principle of liquid displacement. The results were automatically recorded and displayed by integrated embedded data acquisition system. Each digital pulse corresponded to the defined volume of gas coming from the equipment.

Two experiments were performed in a batch mode, each of these in a different temperature. In the first one, the mesophilic conditions (37±1°C) were maintained. The R 1.1 reactor (the control) was fed with 0.4 dm^3 of SS, while the R 1.2 reactor was supplied by mixture of the SS (0.4 dm^3) and 5 g of dried BSG as an additional substrate. The second experiment was conducted assuming the analogous schedule; however, the temperature was increased to thermophilic conditions (55±1°C). Therein, the substrates doses were kept unchanged. The R 2.1 reactor was a control one, whereas R 2.2 was supplied by BSG and SS.

2.3 *Analytical methods*

The substrates were analyzed once after delivery to laboratory. In both samples total solids (TS), volatile solids (VS) and pH were measured. Total and volatile solids were determined according to the Standard Methods for the Examination of Water and Wastewater (APHA 2005). The pH values were controlled by means of HQ 40D Hach-Lange multimeter (Hach, Loveland, CO, USA) in compliance with PN-EN 12176.

The biogas composition (CH_4 and CO_2) was determined using a Shimadzu GC 14B gas chromatograph coupled with a thermal conductivity detector fitted with glass packed columns. The parameters used for the analysis were 40°C for the injector and 60°C for the detector. The carrier gas was helium with a flux rate of 40 cm^3·min^{-1}. Peak areas were determined by means of the computer integration program (CHROM-CARD).

3 RESULTS AND DISCUSSION

The average biogas and methane potentials as well as kinetic parameters are listed in Table 2. The obtained results indicated that the co-digestion of BSG and SS under both mesophilic and thermophilic conditions resulted in an improvement in the biogas potential (BP_{21}) and biogas production rate (GPR), as compared to the SS mono-digestion. Considering the first parameter, the enhancements of 19 and 26.3 % under meso- and thermophilic temperatures were found, respectively. Meanwhile, as compared to the SS mono-digestion, the biogas production rate increased by 28.2 and 29.3% under meso- and thermophilic conditions, respectively. Analogously, the enhanced GPR was achieved under thermophilic conditions. The improved GPR under thermophilic conditions was also found for the anaerobic co-digestion of food waste and wheat straw (Shi et al. 2018). Therein, the average values of GPR were 0.33-0.56 and 0.42-0.61 $Ndm^3dm^{-3}d^{-1}$ under meso- and thermophilic conditions, respectively. Considering the co-digestion of SS and organic fraction of municipal solid wastes, after 21 days of digestion the GPR reached 0.1 and 1.0 $Ndm^3dm^{-3}d^{-1}$ in the meso- and thermophilic reactor, respectively (Sosnowski et al. 2003). In this study, the difference between the related values was considerably lower.

Additionally, the higher temperature affected the biogas composition. The increased methane content was observed under thermophilic conditions. Nevertheless, compared to the SS mono-digestion, the drop tendency occurred in BSG presence. Therein, the average values were 70.05% ± 0.49 and 68.52% ± 0.14 in the control and co-digestion run, respectively. This effect might have resulted from the BSG application characterized by a significant amount of carbohydrates (Nielfa et al. 2015). The same trend was observed in the study conducted by Zou et al. (2018), where residual sludge was co-digested with different lignocellulosic wastes. Conversely, the slight increase from 62.07% ± 0.33 (R 1.1) to 62.38% ± 0.13 (R 1.2) was achieved under mesophilic conditions.

The application of BSG as additional substrate resulted in improvement of both methane potential and methane production rate. Accordingly, the increased process temperature led to more beneficial results. Compared to mono-digestion, for the MBP_{21} an enhancement of 19.9% and 23.7% occurred under meso- and thermophilic temperatures, respectively. Regarding the MPR, the average values increased by 29.5 and 28.8% in R 1.2 and R 2.2, respectively.

The observed improvements in the co-digestion runs for both temperature regimes might be attributed to the introduction of BSG, which eliminated the deficits of SS mono-digestion. This co-substrate supplemented the feedstock in organic matter as well as the vitamins, mineral salts and amino acids that enhanced the methanogens activity. Moreover, BSG ensured a necessary buffer capacity and improved the C/N ratio in the feedstock. Compared to other wastes, a significant methane and biogas potential of BSG also contributed to achieving such beneficial results in terms of examined parameters. An analogous tendency was found in the study conducted by Zou et al. (2018) who co-digested residual sludge and different lignocellulosic wastes

Table 2. The biogas/methane potential and biogas/methane production rate under meso- and thermophilic conditions (the average value and standard deviation are given).

		SS R 1.1	SS + BSG R 1.2	SS R 2.1	SS + BSG R 2.2
Parameter	Unit	mesophilic conditions*		thermophilic conditions	
BP_{21}	Ndm^3g^{-1} VS	0.405 ± 0.019	0.482 ± 0.036	0.453 ± 0.021	0.572 ± 0.043
GPR	$Ndm^3dm^{-3}d^{-1}$	0.71 ± 0.022	0.91 ± 0.048	0.75 ± 0.023	0.97 ± 0.051
MBP_{21}	Ndm^3g^{-1} VS	0.251 ± 0.011	0.301 ± 0.024	0.317 ± 0.014	0.392 ± 0.032
MPR	$Ndm^3dm^{-3}d^{-1}$	0.44 ± 0.013	0.57 ± 0.023	0.52 ± 0.016	0.67 ± 0.027

* Lebiocka et al., 2018

(greening waste, herbs waste and sugarcane bagasse waste). Therein, more favorable values of methane potential were noticed accompanying the highest doses of co-substrates (50% v/v) than in the present study. The average values of the experimental methane potential varied between 0.307-0.345 $dm^3g^{-1}VS$. The authors also attributed this effect to high carbohydrate fraction and more balanced C/N ratio in the co-digestion runs resulting from the co-substrate addition.

The previous studies have shown that anaerobic digestion of various lignocellulosic materials, including BSG, resulted in poor stability related to inhibition most likely induced by the presence of phenolic compounds, which constitute main by-products of lignin decomposition (Sežun et al. 2011, Panjičko et al. 2017). Moreover, the presented studies have also indicated that the anaerobic degradation of lignin is slow and ineffective, because of its heterogeneous chemical structure which is recalcitrant for most bacteria (Fernandes et al. 2009, Zhong et al. 2011). However, the recent studies have suggested that the appropriately selected co-substrate and process conditions could improve the lignin degradation (Čater et al. 2015, Gonçalves et al. 2016, Zou et al. 2018). This finding for BSG was confirmed in the co-digestion with cow dung and pig manure (Poulsen et al. 2017), cattle dung (Tewelde et al. 2012), Jerusalem artichoke (Malakhova et al. 2015), and monoazo dye and other co-substrates (glucose and sodium acetate) (Gonçalves et al. 2015). In the present study, the observed enhancements in BSG attendance indicated that this synergistic effect was achieved in terms of SS.

Interestingly, the presented results showed that the increased temperature, i.e. thermophilic conditions, affected the co-digestion process favorably. Generally, these conditions are more susceptible to inhibition than mesophilic one (Labatut et al. 2014, Gebreeyessus & Jenicek, 2016). This fact should be particularly considered using a co-substrate that could potentially generate toxic intermediates. Though, in the present research, a stable process performance and improved biogas production were observed. The several studies of anaerobic digestion using lignocellulosic materials have also confirmed that thermophilic conditions could enhance their degradation through an increased microbial hydrolysis rate. It is widely known that hydrolysis constitutes a main limiting step in an effective biogas production (Shi et al. 2013, Hu et al. 2017, Neshat et al. 2017, David et al. 2018). Additionally, Shi et al. (2018) indicated that the population of lignocellulose-degrading microorganisms was improved under the thermophilic regime. According to the authors, this observation could explain the better digester performance at the higher temperature. Thus far, this favorable effect has been reported in terms of anaerobic co-digestion of food waste and manures with different lignocellulosic wastes, e.g. BSG, wheat straw paper wastes, corn stover, bamboo and saw dust (Das Ghatak & Mahanta, 2014, Malakhova et al. 2015, Shi et al. 2018, David et al. 2018). Therefore, these conditions should be recommended for the lignocellulosic biomass used as a feedstock.

The average cumulative biogas production profiles for both experiments are presented in Figure 1. Interestingly, the profiles for SS mono-digestion differed as compared to the co-digestion experiments, and two phases of biogas production revealed, which was especially visible under thermophilic conditions (Figure 1b). Each of them seems to have different kinetics which requires validation in the further studies.

As is shown in Figure 1 a and b, a rapid initial biogas production was observed for both temperature regimes. Under mesophilic conditions, this phase lasted 12 days, while under thermophilic conditions, this time was shortened to 6 days. Under mesophilic conditions in the initial stage (5 days) the biogas production was almost equal for both mono- and co-digestion run (Figure 1a). After this time an increase was observed in the presence of a co-substrate. This effect might be contributed to a complex structure of BSG that require prolonged degradation time. In turn, under thermophilic temperatures (Figure 1b), this phase was shortened to 2 days. After this period, the biogas production was enhanced as compared to the SS mono-digestion. This observation also confirmed the advantage of these conditions on the anaerobic digestion of lignocellulosic materials. In this case, the increased temperature accelerated the hydrolysis stage, ensuring more accessible compounds for microorganisms.

Meanwhile, as compared to the mesophilic temperatures, the biogas production under thermophilic conditions started very slowly (see the curve course in the range from 0 to 4 day), then it accelerated and exceeded the production for the mesophilic experiment (Figure 2).

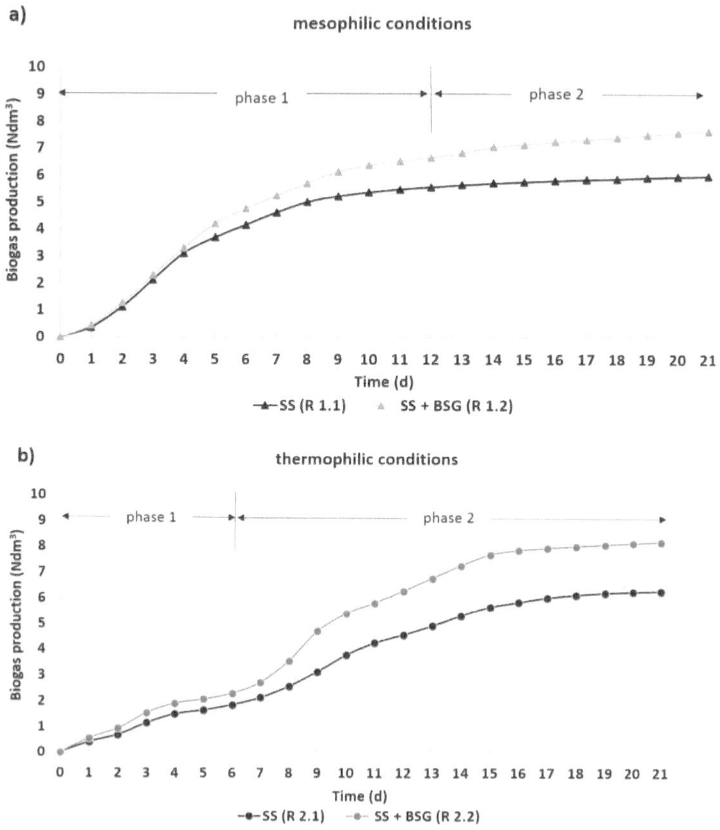

Figure 1. Cumulative biogas production under meso- (a) and thermophilic (b) conditions divided into phases (the average values are presented).

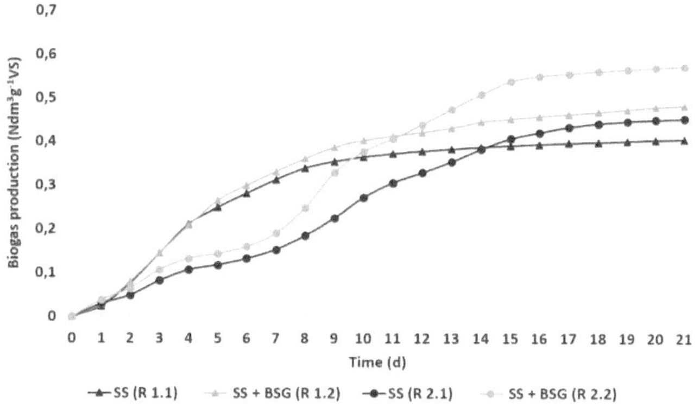

Figure 2. Comparison of cumulative biogas production under meso- and thermophilic conditions (the average values of daily biogas production per g of VS).

For mono-digestion this effect was found after 14 d, while for co-digestion, this time was shortened to 11 d. The analogous observation was found in the study conducted by Sosnowski et al. (2003) who co-digested SS and organic fraction of municipal solid wastes under both temperature regimes.

4 CONCLUSIONS

The result indicated that the application of BSG as an additional substrate improved the biogas/ methane potential and kinetic parameters as compared to the SS mono-digestion. This effect was most likely attributed to the supplementation of the feedstock with BSG that provided a significant amount of organic matter, high alkalinity and various micro- and macro-elements, crucial in the anaerobic digestion process. The increased temperature accelerated the biochemical reaction rates, ensuring more beneficial results in terms of all analyzed parameters. However, while considering the implementation of this strategy to exploiting digesters, the energy efficiency should be evaluated due to the higher energy requirements under thermophilic conditions.

REFERENCES

Čater, M., Fanedl, L., Malovrh, Š. & Marinšek Logar, R. 2015. Biogas production from brewery spent grain enhanced by bioaugmentation with hydrolytic anaerobic bacteria. *Bioresource Technology* 186: 261–26.

Das Ghatak, M. & Mahanta, P. 2014. Effect of Temperature on Anaerobic Co-digestion of cattle dung with Lignocellulosic Biomass. *Journal of Advanced Engineering Research* 1: 1–7.

David, A, Govil, T., Tripathi, A. K., McGeary, J., Farrar, K. & Sani R.K. 2018. Thermophilic Anaerobic Digestion: Enhanced and Sustainable Methane Production from Co-Digestion of Food and Lignocellulosic Wastes. *Energies* 11(8): 1–13.

Fernandes, T.V., Klaasse Bos, G.J., Zeeman, G., Sanders, J.P.M. & van Lier, J.B. 2009 Effects of thermo-chemical pre-treatment on anaerobic biodegradability and hydrolysis of lignocellulosic biomass. *Bioresource Technology* 100: 2575–2579.

Gebreeyessus, G.D. & Jenicek, P. 2016. Thermophilic versus Mesophilic Anaerobic Digestion of Sewage Sludge: A Comparative Review. *Bioengineering (Basel)* 3 (2).

Giuliano, A., Bolzonella, D., Pavan, P., Cavinato, C.& Cecchi, F. 2013. Co-digestion of livestock effluents, energy crops and agro-waste: Feeding and process optimization in mesophilic and thermophilic conditions. *Bioresource Technology* 128; 612–618.

Gonçalves, I.C., Fonsec, A., Morao, A.M., Pinheirob, H.M., Duarte A.P.& Ferra, M.I.A. 2015. Evaluation of anaerobic co-digestion of spent brewery grains and an azo dye, *Renewable Energy* 74: 489–496.

Hu, J., Jing, Y., Zhang, Q., Guo, J.& Lee D.J. 2017. Enzyme hydrolysis kinetics of microgrinded maize straws. *Bioresource Technology* 240: 177–180.

Kainthola, J., Kalamdhad, A.S. & Goud, V.V. 2019. A review on enhanced biogas production from anaerobic digestion of lignocellulosic biomass by different enhancement techniques. *Process Biochemistry* 84: 81–90.

Labatut, R.A., Angenent, L.T. & Scott, N.R. 2014. Conventional mesophilic vs. thermophilic anaerobic digestion: a trade-off between performance and stability? *Water Research* 53: 249–258.

Lebiocka, M., Montusiewicz, A. & Bis, M. 2018. Influence of milling on the effects of co-digestion of brewery spent grain and sewage sludge. In Sobczuk, H. & Kowalska, B. (eds.), *Water supply and wastewater disposal:* 94–101, Politechnika Lubelska, Poland.

Malakhova, D.V., Egorova, M.A., Prokudina, L.I., Netrusov, A.I. & Tsavkelova, E.A. 2015. The biotransformation of brewer's spent grain into biogas by anaerobic microbial communities. *World Journal of Microbiology and Biotechnology* 31: 2015–2023.

Mata-Alvarez, J., Dosta, J., Romero-Güiza, M.S., Fonoll, X., Peces, M. & Astals, S. 2014. A critical review on anaerobic co-digestion achievements between 2010 and 2013. *Renewable and Sustainable Energy Reviews* 36: 412–427.

Moset, V., Poulsen, M., Wahid, R., Højberg, O. & Møller, H.B. 2015. Mesophilic versus thermophilic anaerobic digestion of cattle manure: methane productivity and microbial ecology. *Microb. Biotechnol.* 8: 787–800.

Mussatto, S.I., Dragone G. & Roberto I.C. 2006. Brewers' spent grain: generation, characteristics and potential applications. *Journal of Cereal Science* 43(1): 1–14.

Neshat, S.A., Mohammadi, M., Najafpour, G. D.& Lahijani P. 2017. Anaerobic co-digestion of animal manures and lignocellulosic residues as a potent approach for sustainable biogas production. *Renewable and Sustainable Energy Reviews* 79: 308–322.

Nielfa, A., Cano, R., Perez, A. & Fdez-Polanco, M. 2015. Co-digestion of municipal sewage sludge and solid waste: modelling of carbohydrate, lipid and protein content influence. *Waste Management & Research* 33: 241–249.

Panjičko, M., Zupančič, G.D., Fanedl, L., Logar, R. M. Tišma, M. & Zelić, B. 2017. Biogas production from brewery spent grain as a mono-substrate in a two-stage process composed of solid-state anaerobic digestion. *Journal of Cleaner Production* 166: 519–529.

Poulsen T.G., Adelard L. & Wells M. 2017. Improvement in CH_4/CO_2 ratio and CH_4 yield as related to biomass mix composition during anaerobic co-digestion. *Waste Management* 61: 179–187.

Retfalvi, T., Tukacs-Hajos A. & Szabo P. 2013. Effects of artificial overdosing of p-cresol and phenylacetic acid on the anaerobic fermentation of sugar beet pulp. *International Biodeterioration & Biodegradation* 83: 112–118.

Sawatdeenarunat, C., Surendra, K.C., Takara, D., Oechsner, H. & Khanal, S.K. 2015. Anaerobic digestion of lignocellulosic biomass: challenges and opportunities. *Bioresource Technology* 178: 178–186.

Sežun, M., Grilc, V., Zupančič, G.D. & Marinšek-Logar, R. 2011. Anaerobic digestion of brewery spent grain in a semi-continuous bioreactor: inhibition by phenolic degradation products. *Acta Chimica Slovenica*, 58(1): 158–66.

Shi, J., Wang, Z, Stiverson, J.A., Yu, Z. & Li Y. 2013. Reactor performance and microbial community dynamics during solid-state anaerobic digestion of corn stover at mesophilic and thermophilic conditions. *Bioresource Technology* 136: 574–81.

Shi, X., Guo, X., Zuo, J., Wang, Y. & Zhang, M. 2018. A comparative study of thermophilic and mesophilic anaerobic co-digestion of food waste and wheat straw: Process stability and microbial community structure shifts. *Waste Management* 77: 261–269.

Soheil A. Neshata S.A., Mohammadi, M., Najafpoura, G.D. & Lahijani, P. 2013. Anaerobic co-digestion of animal manures and lignocellulosic residues as a potent approach for sustainable biogas production. *Bioresource Technology* 136: 574–81.

Sosnowski, P., Wieczorek, A. & Ledakowicz, S. 2003. Anaerobic co-digestion of sewage sludge and organic fraction of municipal solid wastes. *Advances in Environmental Research* 7(3): 609–616.

Tewelde, S., Eyalarasan, K., Radhamani, R. & Karthikeyan, K. 2012. Biogas production from co-digestion of brewery wastes [BW] and cattle dung [CD]. *International Journal of Latest Trends in Agriculture and Food Sciences* 2: 90–93.

Yang, L., Xu, F., Ge, X. & Li, Y. 2015. Challenges and strategies for solid-state anaerobic digestion of lignocellulosic biomass. *Renewable and Sustainable Energy Reviews* 44: 824–834.

Zhang, C., Su, H., Baeyens, J. & Tan, T. 2014. Reviewing the anaerobic digestion of food waste for biogas production, *Renewable and Sustainable Energy Reviews* 38: 383–392.

Zhong, W., Zhang Z., Luo, Y., Sun, Sh., Qiao, W. & Xiao, M. 2011. Effect of biological pretreatments in enhancing corn straw biogas production. *Bioresource Technology* 102 (24): 11177–11182.

Zou, H., Chen, Y., Shi, J., Zhao, T., Yu, Q., Yu, Sh., Shi, D., Chai, H., Gu, L., He, Q. & Ai, H. 2018. Mesophilic anaerobic co-digestion of residual sludge with different lignocellulosic wastes in the batch digester. *Bioresource Technology* 268: 371–381.

The Role of Agriculture in Climate Change Mitigation – Pawłowski, Litwińczuk & Zhou (eds)
© 2020 Taylor & Francis Group, London, ISBN 978-0-367-43372-7

Author Index